STUDENT SOLUTIONS MANUAL

TO ACCOMPANY

UNDERSTANDABLE STATISTICS
EIGHTH EDITION
BRASE/BRASE

STUDENT SOLUTIONS MANUAL

TO ACCOMPANY

UNDERSTANDABLE STATISTICS
EIGHTH EDITION
BRASE/BRASE

Charles Henry Brase
Regis University

Corrinne Pellillo Brase
Arapahoe Community College

Laurel Tech Integrated Publishing Services

HOUGHTON MIFFLIN COMPANY BOSTON NEW YORK

Sponsoring Editor: Lauren Schultz
Development Editor: David George
Editorial Associate: Kasey McGarrigle
Manufacturing Coordinator: Karen Banks
Senior Marketing Manager: Ben Rivera
Senior Project Editor: Nancy Blodget

Printed in the U.S.A.

ISBN: 0-618-50159-2

1 2 3 4 5 6 7 8 9-EB-09 08 07 06 05

Contents

Chapter 1 Getting Started

Section 1.1

1. **(a)** The variable is the response regarding frequency of eating at fast-food restaurants.

(b) The variable is qualitative. The categories are the number of times one eats in fast-food restaurants.

(c) The implied population is responses for all adults in the U.S.

3. **(a)** The variable is student/faculty ratio at colleges.

(b) The variable is quantitative because arithmetic operations can be applied to the ratios.

(c) The implied population is student/faculty ratio at all colleges in the nation.

5. **(a)** The variable is the nitrogen concentration (mg nitrogen/l water).

(b) The variable is quantitative because arithmetic operations can be applied to the time intervals.

(c) The implied population is the nitrogen concentration (mg nitrogen/l water) in the entire lake.

7. **(a)** *Length of time to complete an exam* is a ratio level of measurement. The data may be arranged in order, differences and ratios are meaningful, and a time of 0 is the starting point for all measurements.

(b) *Time of first class* is an interval level of measurement. The data may be arranged in order and differences are meaningful.

(c) *Major field of study* is a nominal level of measurement. The data consists of names only.

(d) *Course evaluation scale* is an ordinal level of measurement. The data may be arranged in order.

(e) *Score on last exam* is a ratio level of measurement. The data may be arranged in order, differences and ratios are meaningful, and a score of 0 is the starting point for all measurements.

(f) *Age of student* is a ratio level of measurement. The data may be arranged in order, differences and ratios are meaningful, and an age of 0 is the starting point for all measurements.

9. **(a)** *Species of fish* is a nominal level of measurement. Data consist of names only.

(b) *Cost of rod and reel* is a ratio level of measurement. The data may be arranged in order, differences and ratios are meaningful, and a cost of 0 is the starting point for all measurements.

(c) *Time of return home* is an interval level of measurement. The data may be arranged in order and differences are meaningful.

(d) *Guidebook rating* is an ordinal level of measurement. Data may be arranged in order.

(e) *Number of fish* caught is a ratio level of measurement. The data may be arranged in order, differences and ratios are meaningful, and 0 fish caught is the starting point for all measurements.

(f) *Temperature of the water* is an interval level of measurement. The data may be arranged in order and differences are meaningful.

Section 1.2

1. Essay

3. Answers vary. Use groups of 4 digits.

5. **(a)** Assign a distinct number to each subject. Then use a random number table. Group assignment methods vary.

(b) Repeat part (a) for 22 subjects.

(c) Answers vary.

7. (a) Yes, it is appropriate that the same number appears more than once because the outcome of a die roll can repeat. The outcome of the 4th roll is 2.

(b) No, we do not expect the same sequence because the process is random.

9. (a) Reasons may vary. For instance, the first four students may make a special effort to get to class on time.

(b) Reasons may vary. For instance, four students who come in late might all be nursing students enrolled in an anatomy and physiology class that meets the hour before in a far-away building. They may be more motivated than other students to complete a degree requirement.

(c) Reasons may vary. For instance, four students sitting in the back row might be less inclined to participate in class discussions.

(d) Reasons may vary. For instance, the tallest students might all be male.

11. In all cases, assign distinct numbers to the items, and use a random-number table.

13. Answers vary. Use single digits with correct answer placed in corresponding position.

15. (a) This technique is simple random sampling. Every sample of size n from the population has an equal chance of being selected and every member of the population has an equal chance of being included in the sample.

(b) This technique is cluster sampling. The state, Hawaii, is divided into regions using, say, the first 3 digits of the Zip code. Within each region a random sample of 10 Zip code areas is selected using, say, all 5 digits of the Zip code. Then, within each selected Zip codes, <u>all</u> businesses are surveyed. The sampling units, defined by 5 digit Zip codes, are clusters of businesses, and within each selected Zip code, the benefits package the businesses offer their employees differs business to business.

(c) This technique is convenience sampling. This technique uses results or data that are conveniently and readily obtained.

(d) This technique is systematic sampling. Every k^{th} element is included in the sample.

(e) This technique is stratified sampling. The population was divided into strata (10 business types), then a simple random sample was drawn from each stratum.

Section 1.3

1. (a) This is an observational study because observations and measurements of individuals are conducted in a way that doesn't change the response or the variable being measured.

(b) This is an experiment because a treatment is deliberately imposed on the individuals in order to observe a possible change in the response or variable being measured.

(c) This is an experiment because a treatment is deliberately imposed on the individuals in order to observe a possible change in the response or variable being measured.

(d) This is an observational study because observations and measurements of individuals are conducted in a way that doesn't change the response or the variable being measured.

3. (a) Sampling was used because measurements from a representative part of the population were used.

(b) A simulation was used because computer programs that mimic actual flight were used.

(c) A census was used because data for <u>all</u> scores are available.

(d) An experiment was used. A treatment is deliberately imposed on the individuals in order to observe change in the response or variable being measured.

5. (a) Use random selection to pick 10 calves to inoculate. Then test all calves to see if there is a difference in resistance to infection between the two groups. There is no placebo being used.

(b) Use random selection to pick 9 schools to visit. Then survey all the schools to see if there is a difference in views between the two groups. There is no placebo being used.

(c) Use random selection to pick 40 volunteers for skin patch with drug. Then survey all volunteers to see if a difference exists between the two groups. A placebo for the remaining 35 volunteers in the second group is used.

Chapter 1 Review

1. Answers vary.

3. Essay

5. In the random number table use groups of 2 digits. Select the first six distinct groups of 2 digits that fall in the range from 01 to 42. Choices vary according to the starting place in the random number table.

7. **(a)** This is an observational study because observations and measurements of individuals are conducted in a way that doesn't change the response or the variable being measured.

(b) This is an experiment because a treatment is deliberately imposed on the individuals in order to observe a possible change in the response or variable being measured.

9. This is a good problem for class discussion. Some items such as age and grade point average might be sensitive information. You could ask the class to design a data form that can be filled out anonymously. Other issues to discuss involve the accuracy and honesty of the responses.

11. **(a)** This is an experiment, since a treatment is imposed on one colony.

(b) The control group receives normal daylight/darkness condition. The treatment group has light 24 hours per day.

(c) The number of fireflies living at the end of 72 hours.

(d) The variable is a ratio level of measurement. The data may be arranged in order, differences and ratios are meaningful, and the number 0 fireflies is the starting point for all measurements.

Chapter 2 Organizing Data

Section 2.1

1. Highest Level of Education and Average Annual Household Income (in thousands of dollars)

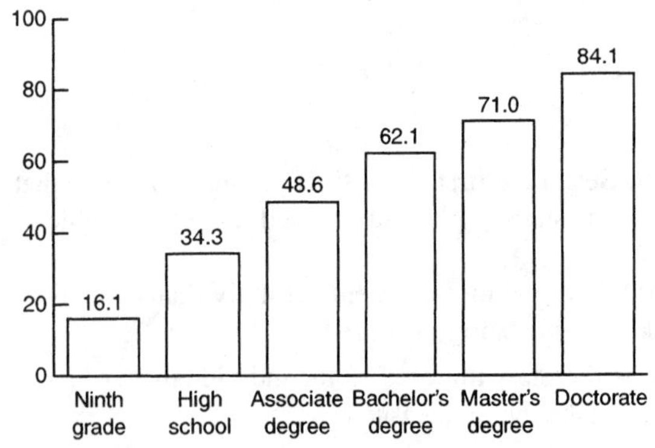

3. Pareto Chart
Annual Harvest for Commercial Fishing in the Gulf of Alaska (in thousand metric tons)

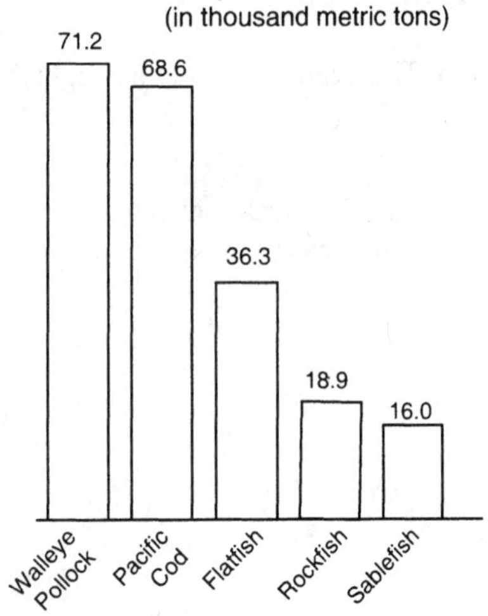

5.

Hiding place	Percentage	Number of Degrees
In the closet	68%	$68\% \times 360° \approx 245°$
Under the bed	23%	$23\% \times 360° \approx 83°$
In the bathtub	6%	$6\% \times 360° \approx 22°$
In the freezer	3%	$3\% \times 360° \approx 11°$
Total	100%	361°*

*Total does not add to 360° due to rounding.

Where We Hide the Mess

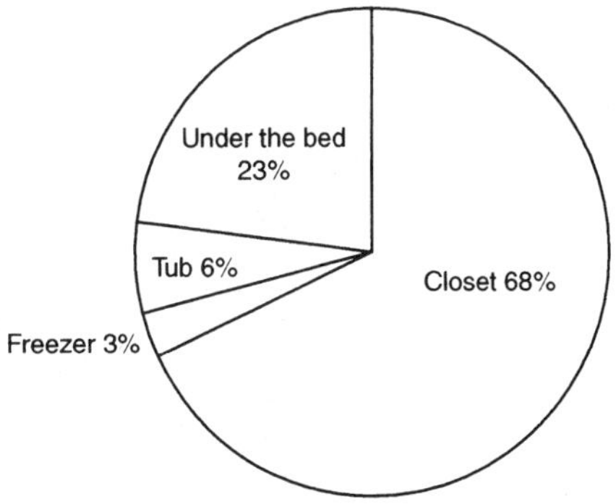

7.

Professional Activity	Percentage	Number of Degrees
Teaching	51%	$51\% \times 360° \approx 184°$
Research	16%	$16\% \times 360° \approx 58°$
Professional growth	5%	$5\% \times 360° = 18°$
Community service	11%	$11\% \times 360° \approx 40°$
Service to the college	11%	$11\% \times 360° \approx 40°$
Consulting outside the college	6%	$6\% \times 360° \approx 22°$
Total	100%	362°*

* Total does not add to 360° due to rounding.

How College Professors Spend Time

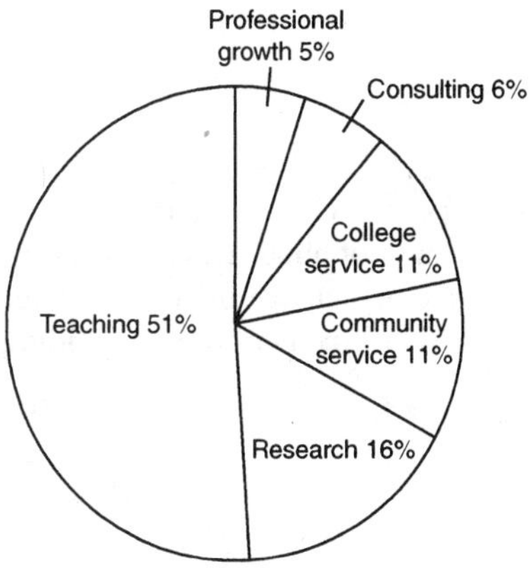

9. (a)

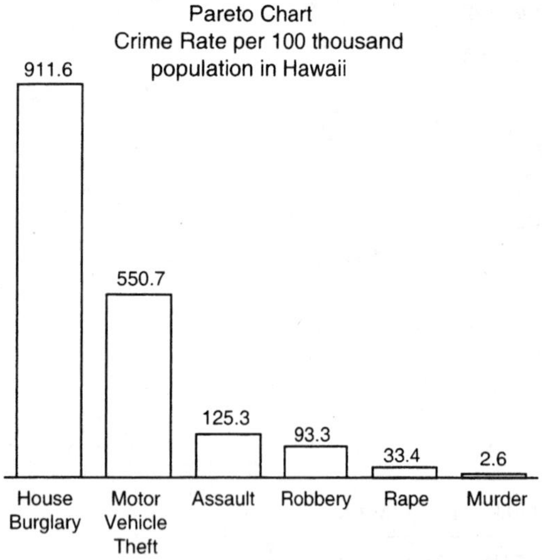

(b) No. In a circle graph, wedges of a circle display proportional parts of the total population that share a common characteristic. The crimes that are listed are not all possible forms. Other forms of crime such as arson are not included in the information. Also, if two or more crimes occur together, the circle graph is not the correct display to choose.

11. Elevation of Pyramid Lake Surface—Time-Series Graph

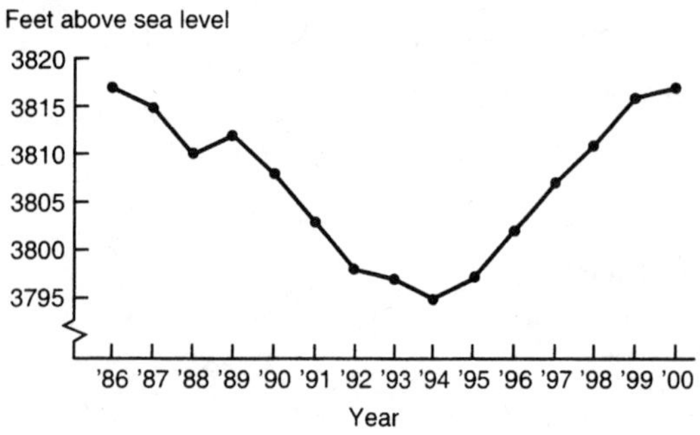

13. Both stocks ended up for the 6-month period. Coca-Cola ranged from a high of about \$48 to a low of about \$37. McDonald's ranged from a high of about \$23 to a low of about \$12. For this 6-month time period, we can calculate the approximate percentage increase in price by finding the difference in the high values (for the last time period and the first time period) and dividing by the first time period high value.

Coca-Cola increased $\frac{45-44}{44} \approx 2\%$

McDonald's increased $\frac{21-15}{15} = 40\%$

Since the time-series graph for McDonald's shows less sharp up and down fluctuations. This indicates its price is less volatile than that of Coca-Cola.

Section 2.2

1. (a) largest data value = 360

smallest data value = 236

number of classes specified = 5

class width $= \dfrac{360-236}{5} = 24.8$, increased to next whole number, 25

(b) The lower class limit of the first class is the smallest value, 236.

The lower class limit of the next class is the previous class's lower class limit plus the class width; for the second class, this is 236 + 25 = 261.

The upper class limit is one value less than lower class limit of the next class; for the first class, the upper class limit is 261 – 1 = 260.

The class boundaries are the halfway points between (i.e., the average of) the (adjacent) upper class limit of one class and the lower class limit of the next class. The lower class boundary of the first class is the lower class limit minus one-half unit. The upper class boundary for the last class is the upper class limit plus one-half unit. For the first class, the class boundaries are $236-\frac{1}{2}=235.5$ and $\frac{260+261}{2}=260.5$. For the last class, the class boundaries are $\frac{335+336}{2}=335.5$ and $360+\frac{1}{2}=360.5$.

The class mark or midpoint is the average of the class limits for that class. For the first class, the midpoint is $\frac{236+260}{2}=248$.

The class frequency is the number of data values that belong to that class; call this value f.

The relative frequency of a class is the class frequency, f, divided by the total number of data values, i.e., the overall sample size, n.

For the first class, $f = 4$, $n = 57$, and the relative frequency is $f/n = \frac{4}{57} \approx 0.07$.

The cumulative frequency of a class is the sum of the frequencies for all previous classes, plus the frequency of that class. For the first and second classes, the class cumulative frequencies are 4 and 4 + 9 = 13, respectively.

Class Limits	Boundaries	Midpoint	Frequency	Relative Frequency	Cumulative Frequency
236–260	235.5–260.5	248	4	0.07	4
261–285	260.5–285.5	273	9	0.16	13
286–310	285.5–310.5	298	25	0.44	38
311–335	310.5–335.5	323	16	0.28	54
336–360	335.5–360.5	348	3	0.05	57

(c) The histogram plots the class frequencies on the y-axis and the class boundaries on the x-axis. Since adjacent classes share boundary values, the bars touch each other. [Alternatively, the bars may be centered over the class marks (midpoints).]

(d) A frequency polygon connects the midpoints of each class (shown as a dot in the middle of the top of the histogram bar) with line segments. Place a dot on the x-axis one class width below the midpoint of the first class, and place another dot on the x-axis one class width above the last class's midpoint. Connect these dots to the adjacent midpoint dots with line segments.

(e) The relative frequency histogram is exactly the same shape as the frequency histogram, but the vertical scale is relative frequency, f/n, instead of actual frequency, f.

The following figure shows the histogram, frequency polygon, and relative-frequency histogram for (c), (d), and (e) above, overlaying one another. (Note that two vertical scales are shown.)

Hours to Complete the Iditarod—Histogram, Frequency Polygon, Relative-Frequency Histogram

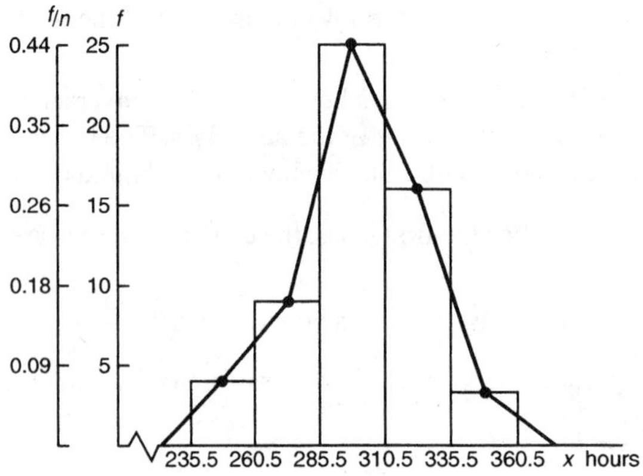

(f) To create the ogive, place a dot on the x-axis at the lower class boundary of the first class and then, for each class, place a dot above the upper class boundary value at the height of the cumulative frequency for the class. Connect the dots with line segments.

Hours to Complete Iditarod—Ogive

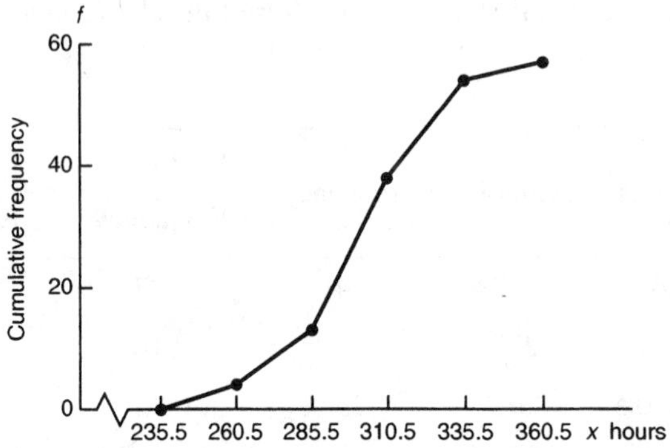

3. (a) largest data value = 59

smallest data value = 1

number of classes specified = 5

class width $= \dfrac{59-1}{5} = 11.6$, increased to next whole number, 12

(b) The lower class limit of the first class is the smallest value, 1.

The lower class limit of the next class is the previous class's lower class limit plus the class width; for the second class, this is 1 + 12 = 13.

The upper class limit is one value less than lower class limit of the next class; for the first class, the upper class limit is 13 – 1 = 12.

The class boundaries are the halfway points between (i.e., the average of) the (adjacent) upper class limit of one class and the lower class limit of the next class. The lower class boundary of the first class is the lower class limit minus one-half unit. The upper class boundary for the last class is the upper class limit plus one-half unit. For the first class, the class boundaries are $1-\frac{1}{2}=0.5$ and $\frac{12+13}{2}=12.5$. For the last class, the class boundaries are $\frac{48+49}{2}=48.5$ and $60+\frac{1}{2}=60.5$.

The class mark or midpoint is the average of the class limits for that class. For the first class, the midpoint is $\frac{1+12}{2}=6.5$.

The class frequency is the number of data values that belong to that class; call this value f.

The relative frequency of a class is the class frequency, f, divided by the total number of data values, i.e., the overall sample size, n.

For the first class, $f = 6$, $n = 42$, and the relative frequency is $f/n = 6/42 \approx 0.14$.

The cumulative frequency of a class is the sum of the frequencies for all previous classes, plus the frequency of that class. For the first and second classes, the class cumulative frequencies are 6 and 6 + 10 = 16, respectively.

Class Limits	Class Boundaries	Midpoint	Frequency	Relative Frequency	Cumulative Frequency
1–12	0.5–12.5	6.5	6	0.14	6
13–24	12.5–24.5	18.5	10	0.24	16
25–36	24.5–36.5	30.5	5	0.12	21
37–48	36.5–48.5	42.5	13	0.31	34
49–60	48.5–60.5	54.5	8	0.19	42

(c) The histogram plots the class frequencies on the y-axis and the class boundaries on the x-axis. Since adjacent classes share boundary values, the bars touch each other. [Alternatively, the bars may be centered over the class marks (midpoints).]

(d) A frequency polygon connects the midpoints of each class (shown as a dot in the middle of the top of the histogram bar) with line segments. Place a dot on the x-axis one class width below the midpoint of the first class, and place another dot on the x-axis one class width above the last class's midpoint. Connect these dots to the adjacent midpoint dots with line segments.

(e) The relative frequency histogram is exactly the same shape as the frequency histogram, but the vertical scale is relative frequency, f/n, instead of actual frequency, f.

The following figure shows the histogram, frequency polygon, and relative-frequency histogram for (c), (d), and (e) above, overlaying one another. (Note that two vertical scales are shown.)

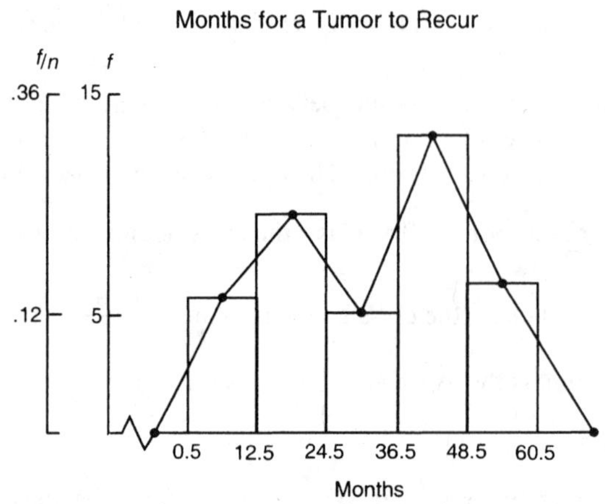

(f) To create the ogive, place a dot on the x-axis at the lower class boundary of the first class and then, for each class, place a dot above the upper class boundary value at the height of the cumulative frequency for the class. Connect the dots with line segments.

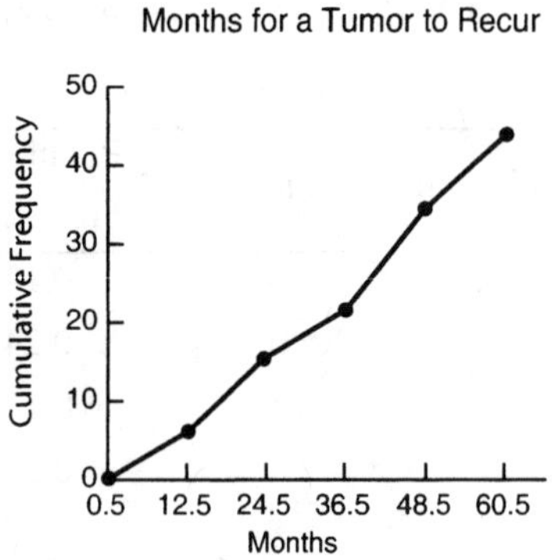

5. (a) largest data value = 52

smallest data value = 10

number of classes specified = 5

class width $= \frac{52-10}{5} = 8.4$, increased to next whole number, 9

(b) The lower class limit of the first class is the smallest value, 10.

The lower class limit of the next class is the previous class's lower class limit plus the class width; for the second class, this is $10 + 9 = 19$.

The upper class limit is one value less than lower class limit of the next class; for the first class, the upper class limit is $19 - 1 = 18$.

The class boundaries are the halfway points between (i.e., the average of) the (adjacent) upper class limit of one class and the lower class limit of the next class. The lower class boundary of the first class is the lower class limit minus one-half unit. The upper class boundary for the last class is the upper class limit plus one-half unit. For the first class, the class boundaries are $10 - \frac{1}{2} = 9.5$ and $\frac{18+19}{2} = 18.5$. For the last class, the class boundaries are $\frac{45+46}{2} = 45.5$ and $54 + \frac{1}{2} = 54.5$.

The class mark or midpoint is the average of the class limits for that class. For the first class, the midpoint is $\frac{10+18}{2} = 14$.

The class frequency is the number of data values that belong to that class; call this value *f*.

The relative frequency of a class is the class frequency, *f*, divided by the total number of data values, i.e., the overall sample size, *n*.

For the first class, $f = 6$, $n = 55$, and the relative frequency is $f/n = 6/55 \approx 0.11$.

The cumulative frequency of a class is the sum of the frequencies for all previous classes, plus the frequency of that class. For the first and second classes, the class cumulative frequencies are 6 and 6 + 26 = 32, respectively.

Highway Fuel Consumption (mpg)

Class Limits	Class Boundaries	Midpoint	Frequency	Relative Frequency	Cumulative Frequency
10–18	9.5–18.5	14	6	0.11	6
19–27	18.5–27.5	23	26	0.47	32
28–36	27.5–36.5	32	20	0.36	52
37–45	36.5–45.5	41	1	0.02	53
46–54	45.5–54.5	50	2	0.04	55

(c) The histogram plots the class frequencies on the *y*-axis and the class boundaries on the *x*-axis. Since adjacent classes share boundary values, the bars touch each other. [Alternatively, the bars may be centered over the class marks (midpoints).]

(d) A frequency polygon connects the midpoints of each class (shown as a dot in the middle of the top of the histogram bar) with line segments. Place a dot on the *x*-axis one class width below the midpoint of the first class, and place another dot on the *x*-axis one class width above the last class's midpoint. Connect these dots to the adjacent midpoint dots with line segments.

(e) The relative frequency histogram is exactly the same shape as the frequency histogram, but the vertical scale is relative frequency, *f/n*, instead of actual frequency, *f*.

The following figure shows the histogram, frequency polygon, and relative-frequency histogram for (c), (d), and (e) above, overlaying one another. (Note that two vertical scales are shown.)

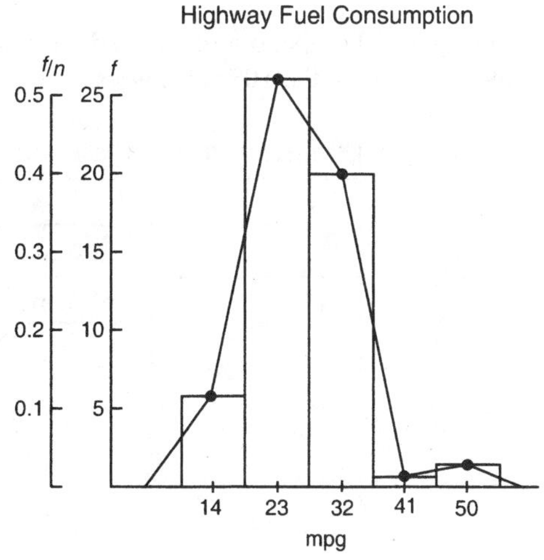

(f) To create the ogive, place a dot on the x-axis at the lower class boundary of the first class and then, for each class, place a dot above the upper class boundary value at the height of the cumulative frequency for the class. Connect the dots with line segments.

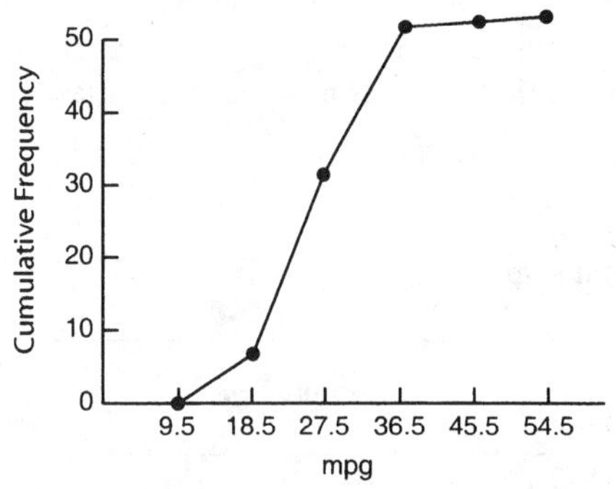

7. (a) The class midpoint is the average of the class limits for that class.

Class Midpoints: 34.5; 44.5; 54.5; 64.5; 74.5; 84.5.

(b) The frequency polygon connects to the x-axis one class width below the smallest midpoint and one class width above the largest midpoint; here the class width is 10, so we have points on the x-axis at $34.5 - 10 = 24.5$ and $84.5 + 10 = 94.5$.

Age of Senators—Frequency Polygons

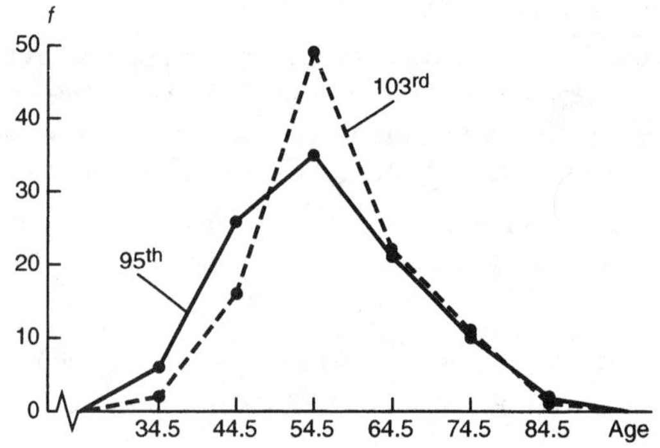

(c) The two polygons have the same general shape, but the dashed polygon is shifted slightly to the right (older ages), so, in general, the members of the 103rd Congress are older.

9. (a)

	Largest value	Smallest value	Class width
Food Companies	11	−3	$\frac{11-(-3)}{5} = 2.8$; use 3
Electronic Companies	16	−6	$\frac{16-(-6)}{5} = 4.4$; use 5

Profit as Percent of Sales—Food Companies

Class	Frequency	Midpoint
−3 to −1	2	−2
0–2	16	1
3–5	10	4
6–8	9	7
9–11	2	10

Profit as Percent of Sales—Electronic Companies

Class	Frequency	Midpoint
−6 to −2	3	−4
−1 to 3	13	1
4–8	20	6
9–13	7	11
14–18	1	16

Profit as a Percent of Sales

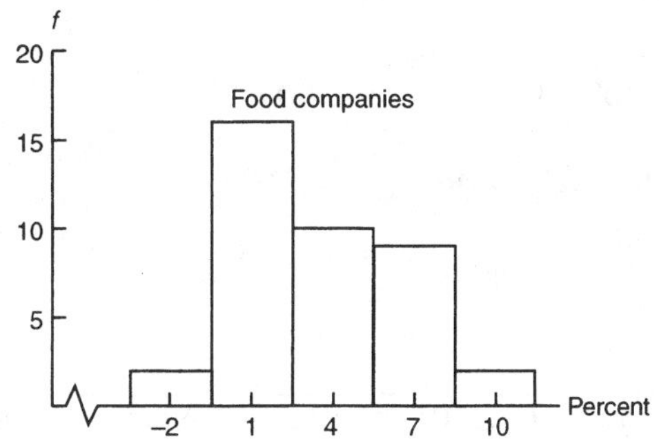

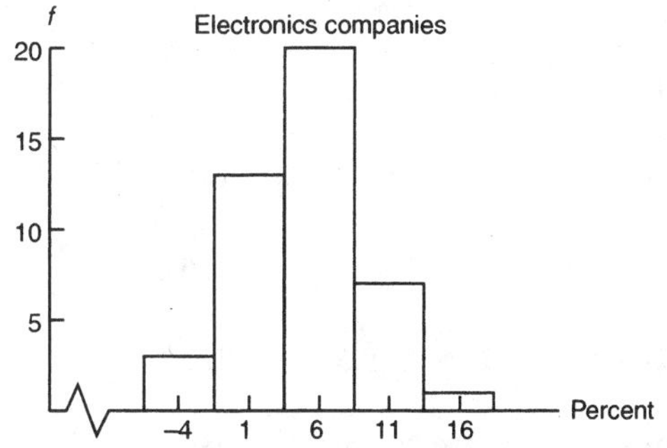

(b) Because the classes and class widths are different for the two company types, it is difficult to compare profits as a percentage of sales. We can notice that for the electronic companies the 16 profits as a percentage of sales extends as high as 18, while for the food companies the highest profit as a percentage of sales is 11. On the other hand, some of the electronic companies also have greater losses than the food companies. Had we made the class limits the same for both company types and overlaid the histograms, it would be easier to compare the data.

11. (a) Since ogives show the cumulative frequency at the upper class boundary, and begin at the point with (x, y) coordinates (lower class boundary of the first class, 0), the numbers on the x-axis are class boundaries. Recall that class boundary values are not values the data can attain. Thus the point marked 85 over the x-value 7.15 means that 85 winning times were less than or equal to 7.15 and, since 7.15 is not a possible data value, 85 winning times were less than 7.15 (i.e., less than 2 minutes 7.15 seconds), Eighty-five of 101 times are less than 7.15, or $\frac{85}{101} \approx 84.2\%$.

(b) Subtract the cumulative frequency at or below 5.15 seconds from the cumulative frequency at or below 11.15 seconds (which includes all the values at or below 5.15 seconds) to get the number of winning times between 5.15 and 11.15 seconds/over two minutes): $100 - 75 = 25$, or $\frac{25}{101} \approx 24.8\%$.

13. (a) Uniform is rectangular, symmetric looks like mirror images on each side of the middle, bimodal has two modes (peaks), and skewed distributions have long tails on one side, and are skewed in the direction of the tail ("skew, few"). (Note that uniform distributions are also symmetric, but "uniform" is more descriptive.)

(a) skewed left, (b) uniform, (c) symmetric, (d) bimodal, (e) skewed right.

(b) Answers vary. Students would probably like (a) since there are many high scores and few low scores. Students would probably dislike (e) since there are few high scores but lots of low scores. (b) is designed to give approximately the same number of As, Bs, etc. (d) has more Bs and Ds, for example. (c) is the way many tests are designed: As and Fs for the exceptionally high and low scores with most students receiving Cs.

15. (a) $2.71 \times 100 = 271$, $1.62 \times 100 = 162, \ldots, 0.70 \times 100 = 70$.

(b) largest value = 282, smallest value = 46

$$\text{class width} = \frac{282-46}{6} \approx 39.3\text{; use 40}$$

Class Limits	Class Boundaries	Midpoint	Frequency
46–85	45.5–85.5	65.5	4
86–125	85.5–125.5	105.5	5
126–165	125.5–165.5	145.5	10
166–205	165.5–205.5	185.5	5
206–245	205.5–245.5	225.5	5
246–285	245.5–285.5	265.5	3

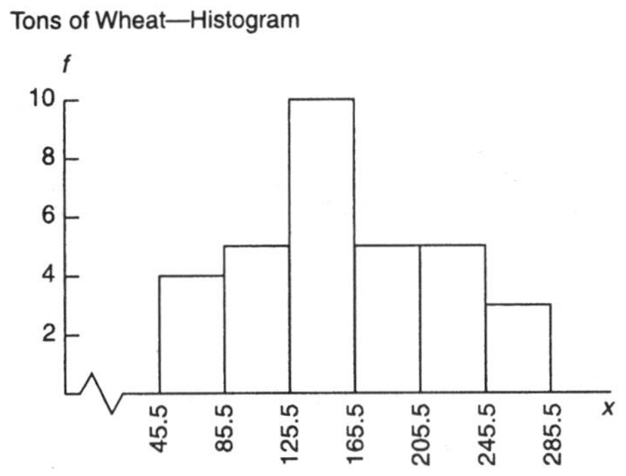

(c) class width is $\frac{40}{100} = 0.40$

Class Limits	Class Boundaries	Midpoint	Frequency
0.46–0.85	0.455–0.855	0.655	4
0.86–1.25	0.855–1.255	1.055	5
1.26–1.65	1.255–1.655	1.455	10
1.66–2.05	1.655–2.055	1.855	5
2.06–2.45	2.055–2.455	2.255	5
2.46–2.85	2.455–2.855	2.655	3

17. **(a)** There is one dot below 600, so 1 state has 600 or fewer licensed drivers per 1000 residents.

(b) 5 values are close to 800; $\frac{5}{51} \approx 0.0980 \approx 9.8\%$

(c) 9 values below 650
37 values between 650 and 750
5 values above 750
From either the counts or the dotplot, the interval from 650 to 750 licensed drivers per 1000 residents has the most "states."

19. The dotplot shows some of the characteristics of the histogram, such as the concentration of most of the data in two peaks, one from 13 to 24 and another from 37 to 48.

However, they are somewhat difficult to compare since the dotplot can be thought of as a histogram with one value, the class mark, i.e., the data value, per class.

Because the definitions of the classes and, therefore, the class widths differ, it is difficult to compare the two figures.

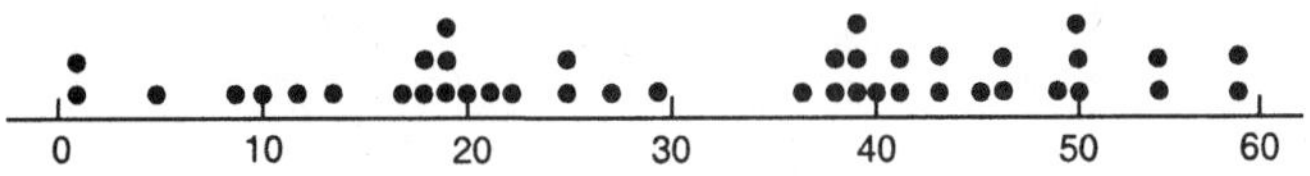

Section 2.3

1. (a) The smallest value is 47 and the largest is 97, so we need stems 4, 5, 6, 7, 8, and 9. Use the tens digit as the stem and the ones digit as the leaf.

Longevity of Cowboys

4	7 = 47 years
4	7
5	2 7 8 8
6	1 6 6 8 8
7	0 2 2 3 3 5 6 7
8	4 4 4 5 6 6 7 9
9	0 1 1 2 3 7

(b) Yes, certainly these cowboys lived long lives, as evidenced by the high frequency of leaves for stems 7, 8, and 9 (i.e., 70-, 80-, and 90-year olds).

3. The longest average length of stay is 11.1 days in North Dakota and the shortest is 5.2 days in Utah. We need stems from 5 to 11. Use the digit(s) to the left of the decimal point as the stem, and the digit to the right as the leaf.

Average Length of Hospital Stay

5	2 = 5.2 days
5	2 3 5 5 6 7
6	0 2 4 6 6 7 7 8 8 8 8 9 9
7	0 0 0 0 0 0 1 1 1 2 2 2 3 3 3 3 4 4 5 5 6 6 8
8	4 5 7
9	4 6 9
10	0 3
11	1

The distribution is skewed right.

5. (a) The longest time during 1961–1980 was 23 minutes (i.e., 2:23) and the shortest time was 9 minutes (2:09). We need stems 0, 1, and 2. We'll use the tens digit as the stem and the ones digit as the leaf, placing leaves 0, 1, 2, 3, and 4 on the first stem and leaves 5, 6, 7, 8, and 9 on the second stem.

Minutes Beyond 2 Hours (1961-1980)

0	9 = 9 minutes past 2 hours
0	9 9
1	0 0 2 3 3
1	5 5 6 6 7 8 8 9
2	0 2 3 3

(b) The longest time during the period 1981–2000 was 14 (2:14), and the shortest was 7 (2:07), so we'll need stems 0 and 1 only.

Minutes Beyond 2 Hours (1981-2000)

0	7 = 7 minutes past 2 hours
0	7 7 7 8 8 8 8 9 9 9 9 9 9 9 9
1	0 0 1 1 4

(c) In more recent times, the winning times have been closer to 2 hours, with all 20 times between 7 and 14 minutes over two hours. In the earlier period, more than half the times (12 or 20) were more than 2 hours and 14 minutes.

7. The largest value is 1.808 arc seconds per century; the smallest is 0.008. These values would be coded as 18|08 and 00|08, respectively. We need stems 00 to 18.

Angular Momentum of Stars

00	14 = 0.014 arc sec/century
00	08 14 38 42 50 57
01	73
02	16 19 51
03	51 69
04	30
05	
06	23 67
07	59 88
08	88
09	
10	24 57
11	69 69
12	60 60
13	
14	38
15	
16	16 60
17	
18	08

There are no large gaps, but 4 small gaps. Interestingly, gaps at 05, 09, and 13 are 4 stem units apart, as in the gap from 13 to 17, except that the gap at 15 falls between 13 and 17. This <u>might</u> indicate a cycle. The *midrange** is the average of the largest and smallest values; here, $\frac{0.008+1.808}{2} = 0.908$. If this value had occurred in the data, it would be shown as 09|08. In a sense, this value locates the "middle" of the data. We can see that more of the data (18/28, or approximately 64%) occurs below the midrange than above it (where $10/28 \approx 36\%$ of the data are located).

*The midrange is often calculated in Exploratory Data Analysis (EDA), which is discussed in the next chapter. It is also used in nonparametrics, which is the topic of Chapter 12.

9. The largest value in the data is 29.8 mg of tar per cigarette smoked, and the smallest value is 1.0. We will need stems from 1 to 29, and we will use the numbers to the right of the decimal point as the leaves.

Milligrams of Tar per Cigarette

1	0 = 1.0 mg tar
1	0
2	
3	
4	1 5
5	
6	
7	3 8
8	0 6 8
9	0
10	
11	4
12	0 4 8
13	7
14	1 5 9
15	0 1 2 8
16	0 6
17	0
29	8

11. The largest value in the data set is 2.03 mg nicotine per cigarette smoked. The smallest value is 0.13. We will need stems 0, 1, and 2. We will use the number to the left of the decimal point as the stem and the first number to the right of the decimal point as the leaf. The number 2 places to the right of the decimal point (the hundredths digit) will be truncated (chopped off; <u>not</u> rounded off).

Milligrams of Nicotine per Cigarette

0	1 = 0.1 milligram
0	1 4 4
0	5 6 6 6 7 7 7 8 8 9 9 9
1	0 0 0 0 0 0 0 1 2
1	
2	0

13. (a) Average salaries in California range from $49,000 to $126,000. Salaries in New York range from $45,000 to $120,000.

(b) New York has a greater number of average salaries in the $60,000 than California, but California has more average salaries than New York in the $70,000 range.

(c) The California data appear to be similar in shape to the New York data, but California's distribution has been shifted up approximately $10,000. It is also heavier in the upper tail and shows no gap in average salaries, unlike New York which has no salaries in the $110,000 range. California has higher average salaries.

Chapter 2 Review

1. (a) Figure 2-1 (a) (in the text) is essentially a bar graph with a "horizontal" axis showing years and a "vertical" axis showing miles per gallon. However, in depicting the data as a highway and showing it in perspective, the ability to correctly compare bar heights visually has been lost. For example, determining what would appear to be the bar heights by measuring from the white line on the road to the edge of the road along a line drawn from the year to its mpg value, we get the bar height for 1983 to be approximately 7/8 inch and the bar height for 1985 to be approximately 1 3/8 inches (i.e., 11/8 inches). Taking the ratio of the given bar heights, we see that the bar for 1985 should be $\frac{27.5}{26} \approx 1.06$ times the length of the 1983 bar. However, the measurements show a ratio of $\frac{\frac{11}{8}}{\frac{7}{8}} = \frac{11}{7} \approx 1.60$, i.e., the 1985 bar is (visually) 1.6 times the length of the 1983 bar. Also, the years are evenly spaced numerically, but the figure shows the more recent years to be more widely spaced due to the use of perspective.

(b) Figure 2-1(b) is a time-series graph, showing the years on the x-axis and miles per gallon on the y-axis. Everything is to scale and not distorted visually by the use of perspective. It is easy to see the mpg standards for each year, and you can also see how fuel economy standards for new cars have changed over the eight years shown (i.e., a steep increase in the early years and a leveling off in the later years).

(c)

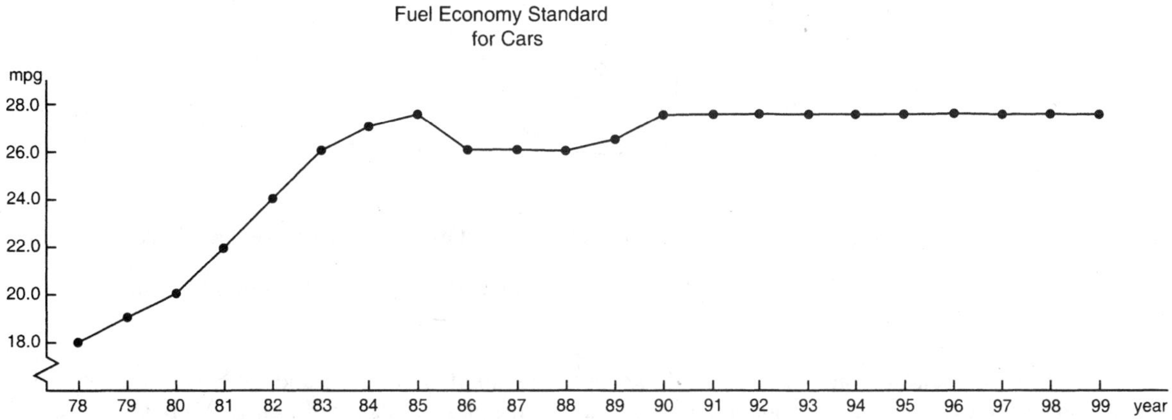

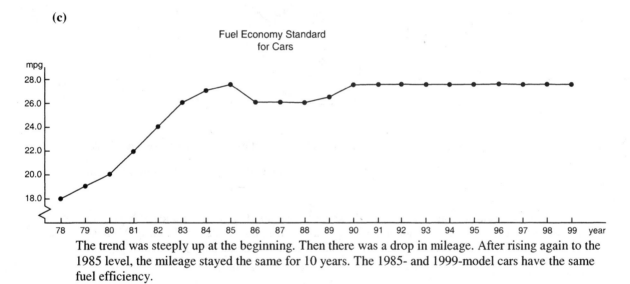

The trend was steeply up at the beginning. Then there was a drop in mileage. After rising again to the 1985 level, the mileage stayed the same for 10 years. The 1985- and 1999-model cars have the same fuel efficiency.

(d)

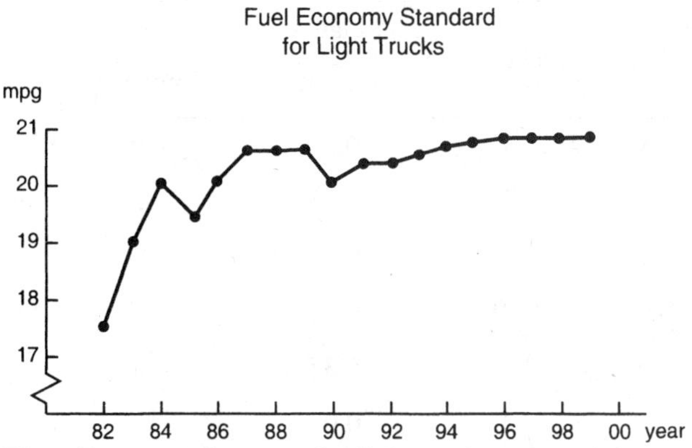

The mileage requirements for light trucks are much lower than for cars, and change much more slowly.

3.

Most Difficult Task	Percentage	Degrees
IRS jargon	43%	$0.43 \times 360° \approx 155°$
Deductions	28%	$0.28 \times 360° \approx 101°$
Right form	10%	$0.10 \times 360° = 36°$
Calculations	8%	$0.08 \times 360° \approx 29°$
Don't know	10%	$0.10 \times 360° = 36°$

Note: Degrees do not total 360° due to rounding.

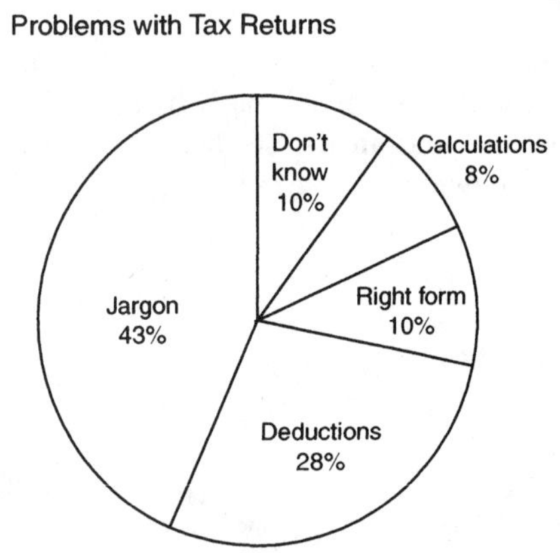

5. (a) The largest value is 142 mm, and the smallest value is 69. For seven classes we need a class width of $\frac{142-69}{7} \approx 10.4$; use 11. The lower class limit of the first class is 69, and the lower class limit of the second class is 69 + 11 = 80.

The class boundaries are the average of the upper class limit of one class and the lower class limit of the next higher class. The midpoint is the average of the class limits for that class. There are 60 data values total so the relative frequency is the class frequency divided by 60.

Class Limits	Class Boundaries	Midpoint	Frequency	Relative Frequency	Cumulative Frequency
69–79	68.5–79.5	74	2	0.03	2
80–90	79.5–90.5	85	3	0.05	5
91–101	90.5–101.5	96	8	0.13	13
102–112	101.5–112.5	107	19	0.32	32
113–123	112.5–123.5	118	22	0.37	54
124–134	123.5–134.5	129	3	0.05	57
135–145	134.5–145.5	140	3	0.05	60

(b) The histogram shows the bars centered over the midpoints of each class.

(c) The frequency polygon begins on the x-axis at the point one class width below the first class midpoint: $74 - 11 = 63$. It connects this point and the other midpoints with line segments. It ends on the x-axis one class width above the last class midpoint: $140 + 11 = 151$.

(d) The frequency histogram and the relative frequency histogram are the same except in the latter, the vertical scale is relative frequency, not frequency.

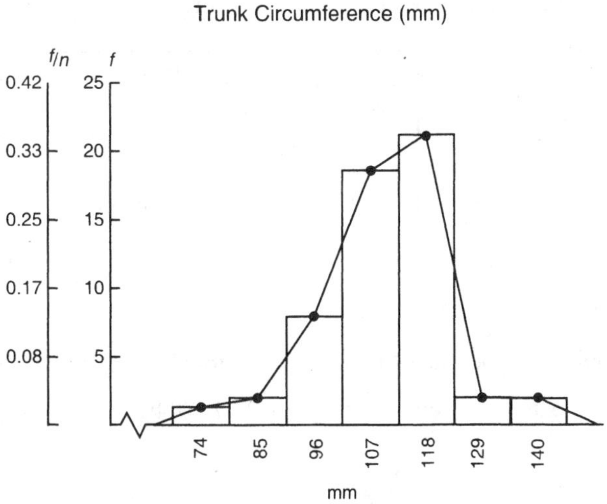

(e) The ogive begins on the x-axis at the lower class boundary and connects dots placed at (x, y) coordinates (upper class boundary, cumulative frequency).

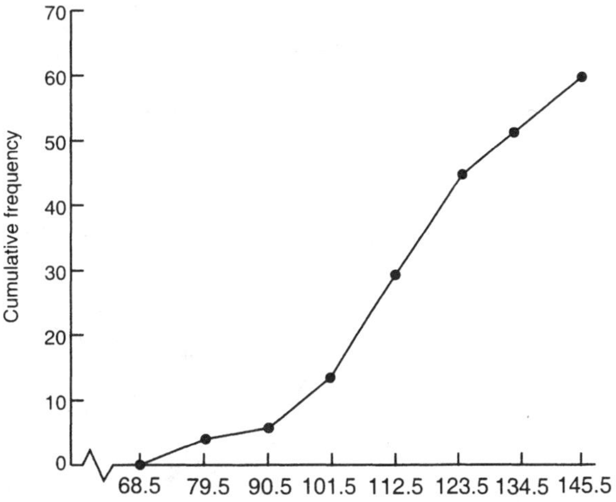

7. (a) To determine the decade which contained the most samples, count both rows (if shown) of leaves; recall leaves 0–4 belong on the first line and 5–9 belong on the second line when two lines per stem are used. The greatest number of leaves is found on stem 124, i.e., the 1240s (the 40s decade in the 1200s), with 40 samples.

(b) The number of samples with tree ring dates 1200 A.D. to 1239 A.D. is $28 + 3 + 19 + 25 = 75$.

(c) The dates of the longest interval with no sample values are 1204 through 1211 A.D. This might mean that for these eight years, the pueblo was unoccupied (thus no new or repaired structures) or that the population remained stable (no new structures needed) or that, say, weather conditions were favorable these years so existing structures didn't need repair. If relatively few new structures were built or repaired during this period, their tree rings might have been missed during sample selection.

9. (a) The age group that is most frequently in the hospital has the highest frequency and, therefore, the highest relative frequency: the age group with boundaries 64.5 and 84.5, enclosing ages 65 to 84.

(b)

Class Limits	Class Boundaries	Relative Frequency	Relative Frequency	Cumulative Relative Frequency (%)
5–24	4.5–24.5	0.16	16%	16%
25–44	24.5–44.5	0.28	28%	44%
45–64	44.5–64.5	0.21	21%	65%
65–84	64.5–84.5	0.35	35%	100%

The percentage of patients older than 44, i.e., from 45 to 84, is 21% + 35% = 56%.

(c) The percentage of patients 44 or younger is 16% + 28% = 44%.

Chapter 3 Averages and Variation

Calculations may vary slightly due to rounding.

Section 3.1

1. Mean $= \bar{x} = \frac{\Sigma x}{n} = \frac{156+161+152+\cdots+157}{12}$

$$= \frac{1876}{12}$$

$$= 156.33$$

The mean is 156.33.

Organize the data from smallest to largest.

144	148	152	153	156	157
157	157	161	161	162	168

To find the median, add the two middle values and divide by 2 since there is an even number of values.

$$\text{Median} = \frac{157+157}{2} = 157$$

The median is 157.

The mode is 157 because it is the value that occurs most frequently.

A gardener in Colorado should look at seed and plant descriptions to determine if the plant can thrive and mature in the designated number of frost-free days. The mean, median, and mode are all close. About half the locations have 157 or fewer frost-free days.

3. Mean $= \bar{x} = \frac{\Sigma x}{n} = \frac{146+152+168+\cdots+144}{14}$

$$= \frac{2342}{14}$$

$$= 167.3$$

The mean is 167.3°F.

Organize the data from smallest to largest.

144	146	152	152	165	168	168
174	178	178	178	179	180	180

To find the median, add the two middle values and divide by 2 since there is an even number of values.

$$\text{Median} = \frac{168+174}{2} = 171$$

The median is 171° F.

The mode is 178° F because it is the value that occurs most frequently.

5. First organize the data from smallest to largest. Then compute the mean, median, and mode.

(a) Upper Canyon

1	1	1	2	3	3	3	3	4	6	9

$$\text{Mean} = \bar{x} = \frac{\Sigma x}{n} = \frac{36}{11} \approx 3.27$$

Median = 3 (middle value)

Mode = 3 (occurs most frequently)

(b) Lower Canyon

0	0	1	1	1	1	2	2	3	6	7	8	13	14

$$\text{Mean} = \bar{x} = \frac{\Sigma x}{n} = \frac{59}{14} \approx 4.21$$

$$\text{Median} = \frac{2+2}{2} = 2$$

Mode = 1 (occurs most frequently)

(c) The mean for the Lower Canyon is greater than that of the Upper Canyon. However, the median and mode for the Lower Canyon are less than those of the Upper Canyon.

(d) 5% of 14 is 0.7 which rounds to 1. So, eliminate one data value from the bottom of the list and one from the top. Then compute the mean of the remaining 12 values.

$$5\% \text{ trimmed mean} = \frac{\Sigma x}{n} = \frac{45}{12} = 3.75$$

Now this value is closer to the Upper Canyon mean.

7. (a) $\text{Mean} = \bar{x} = \frac{\Sigma x}{n} = \frac{23+17+35+\cdots+394}{12}$

$$= \frac{988}{12}$$

$$\approx 82.33$$

The mean is 82.33 arson crimes.

(b) $\text{Median} = \frac{35+39}{2} = 37$

The median is 37 arson crimes.

The median best describes the number of arson crimes for the majority of the areas, since the mean is influenced by the two significantly larger numbers.

(c) $\text{Mean} = \bar{x} = \frac{\Sigma x}{n} = \frac{23+17+35+\cdots+42}{10}$

$$= \frac{309}{10}$$

$$= 30.9$$

The mean is 30.9 arson crimes.

$$\text{Median} = \frac{30+35}{2} = 32.5$$

The median is 32.5 arson crimes.

(d) Without the two extreme values, the mean and the median are closer, and both reflect the number of arson crimes of most of the other areas more accurately. The mean changed quite a bit, while the median did not, a difference that indicates that the mean is more sensitive to the absence or presence of extreme values.

9. (a) $\text{Mean} = \bar{x} = \frac{\Sigma x}{n} = \frac{15+12+\cdots+15}{7} = \frac{102}{7} \approx 14.57$

$\text{Median} = 15$ (middle value)

$\text{Mode} = 15$ (occurs most frequently)

(b) $\text{Mean} = \bar{x} = \frac{\Sigma x}{n} = \frac{102+57+62}{9} = \frac{221}{9} \approx 24.56$

$\text{Median} = 15$ (middle value)

$\text{Mode} = 15$ (occurs most frequently)

(c) The mean is most affected by extreme values.

11. (a) $\text{Mean} = \bar{x} = \frac{\Sigma x}{n} = \frac{2723}{20} = 136.15$

The mean is \$136.15.

The median is \$66.50.

The mode is \$60.

(b) 5% of 20 is 1. Eliminate one data value from the bottom and one from the top of the ordered data. In this case eliminate \$40 and \$500.

$$\text{Mean} = \bar{x} = \frac{\Sigma x}{n} = \frac{2183}{18} \approx 121.28$$

The 5% trimmed mean is \$121.28.

Yes, the trimmed mean more accurately reflects the general level of the daily rental cost, but is still higher than the median.

(c) Median. The low and high prices would be helpful also.

13. (a) Since this data is at the nominal level of measurement, only the mode (if it exists) can be used to summarize the data.

(b) Since this data is at the ratio level of measurement, the mean, median, and mode (if it exists) can be used to summarize the data.

(c) The mode can be used (if it exists). If a 24-hour clock is used, then the data is at the ratio level of measurement, so the mean and median may be used as well.

15. (a) If the largest data value is *replaced* by a larger value, the mean will increase because the sum of the data values will increase, but the number of them will remain the same. The median will not change. The same value will still be in the eighth position when the data are ordered.

(b) If the largest value is replaced by a value that is smaller (but still higher than the median), the mean will decrease because the sum of the data values will decrease. The median will not change. The same value will be in the eighth position in increasing order.

(c) If the largest value is replaced by a value that is smaller than the median, the mean will decrease because the sum of the data values will decrease. The median also will decrease because the former value in the eighth position will move to the ninth position in increasing order. The median will be the new value in the eighth position.

17. $$\text{Weighted average} = \frac{\Sigma xw}{\Sigma w} = \frac{92(0.25)+81(0.225)+93(0.225)+85(0.30)}{1} = 87.65$$

19. $$\text{Weighted average} = \frac{\Sigma xw}{\Sigma w} = \frac{9(2)+7(3)+6(1)+10(4)}{2+3+1+4} = \frac{85}{10} = 8.5$$

21. (a) $$\text{Weighted average} = \frac{\Sigma xw}{\Sigma w} = \frac{5.2(0.15)+3.9(0.20)+3.6(0.40)+2.7(0.25)}{1} \approx 3.7\%$$

(b) $$\text{Weighted average} = \frac{\Sigma xw}{\Sigma w} = \frac{6.9(0.15)+8.9(0.20)+5.2(0.40)+2.6(0.25)}{1} \approx 5.5 \text{ years}$$

(c) $$\text{Total invested} = \$2500+\$4200+\$7850+\$11{,}600 = \$26{,}150$$

Type of Bond	High Yield	Long Term	Int Term	Short Term
% Invested	$\frac{\$2500}{\$26{,}150} = 0.10$	$\frac{\$4200}{\$26{,}150} = 0.16$	$\frac{\$7850}{\$26{,}150} = 0.30$	$\frac{\$11{,}600}{\$26{,}150} = 0.44$

$$\text{Weighted average} = \frac{\Sigma xw}{\Sigma w} = \frac{5.2(0.10)+3.9(0.16)+3.6(0.30)+2.7(0.44)}{1} \approx 3.4\%$$

(d) Weighted average $= \dfrac{\Sigma xw}{\Sigma w}$

$$= \frac{6.9(0.10)+8.9(0.16)+5.2(0.30)+2.6(0.44)}{1}$$

$$\approx 4.8 \text{ years}$$

23. Answers will vary according to data collected.

Section 3.2

1. (a) Range = largest value − smallest value

$$= 30-15$$
$$= 15$$

(b) Use a calculator to verify that $\Sigma x = 110$ and $\Sigma x^2 = 2568$.

(c) Computation formula (sample data) for s^2.

$$s = \sqrt{\frac{\Sigma x^2 - \frac{(\Sigma x)^2}{n}}{n-1}}$$

$$= \sqrt{\frac{2568 - \frac{(110)^2}{5}}{5-1}}$$

$$\approx 6.08$$

$$s^2 = 6.08^2$$
$$\approx 37$$

(d) $\bar{x} = \dfrac{\Sigma x}{n} = \dfrac{110}{5} = 22$

Defining formula (sample data) for s^2.

$$s = \sqrt{\frac{\Sigma(x-\bar{x})^2}{n-1}}$$

$$= \sqrt{\frac{(23-22)^2 + (17-22)^2 + \cdots + (25-22)^2}{5-1}}$$

$$\approx 6.08$$

$$s^2 = 6.08^2$$
$$\approx 37$$

(e) $\mu = 22$

$$\sigma = \sqrt{\frac{\Sigma(x-\mu)^2}{N}}$$
$$= \sqrt{\frac{(23-22)^2 + (17-22)^2 + \cdots + (25-22)^2}{5}}$$
$$\approx 5.44$$

$$\sigma^2 = 5.44^2$$
$$\approx 29.59$$

3. (a) Range = largest value − smallest value
$$= 7.89 - 0.02$$
$$= 7.87$$

(b) Use a calculator to verify that $\Sigma x = 62.11$ and $\Sigma x^2 = 164.23$.

(c) $\overline{x} = \frac{\Sigma x}{n} = \frac{62.11}{50} \approx 1.24$

$$s = \sqrt{\frac{\Sigma x^2 - \frac{(\Sigma x)^2}{n}}{n-1}}$$
$$= \sqrt{\frac{164.23 - \frac{(62.11)^2}{50}}{50-1}}$$
$$\approx 1.333$$
$$\approx 1.33$$

$$s^2 = 1.333^2$$
$$\approx 1.78$$

(d) $CV = \frac{s}{x} \cdot 100 = \frac{1.33}{1.24} \cdot 100 \approx 107\%$

s is 107% of $\overline{x}$.

The standard deviation of the time to failure is just slightly larger than the average time. Since the coefficient of variation can be thought of as a measure of the spread of the data relative to the average of the data, a small *CV* indicates more consistent data whereas a larger *CV* indicates less consistent data.

5. (a) Students verify results with a calculator.

(b) $\bar{x} = \frac{\Sigma x}{n} = \frac{245}{5} = 49$

$$s = \sqrt{\frac{\Sigma x^2 - \frac{(\Sigma x)^2}{n}}{n-1}}$$

$$= \sqrt{\frac{14,755 - \frac{(245)^2}{5}}{5-1}}$$

$$\approx 26.22$$

$s^2 = 26.22^2 \approx 687.49$

(c) $\bar{y} = \frac{\Sigma y}{n} = \frac{224}{5} = 44.8$

$$s = \sqrt{\frac{\Sigma y^2 - \frac{(\Sigma y)^2}{n}}{n-1}}$$

$$= \sqrt{\frac{12,070 - \frac{(224)^2}{5}}{5-1}}$$

$$\approx 22.55$$

$s^2 = 22.55^2 \approx 508.50$

(d) Mallard nest: $CV = \frac{s}{x} \cdot 100 = \frac{26.22}{49} \cdot 100 \approx 53.5\%$

Canada Goose nest: $CV = \frac{s}{y} \cdot 100 = \frac{22.55}{44.8} \cdot 100 \approx 50.3\%$

The *CV* gives the ratio of the standard deviation to the mean. The *CV* for mallard nests is slightly higher.

7. (a) $\bar{x} - 2s = 11.01 - 2(2.17) = 6.67$
$\bar{x} + 2s = 11.01 + 2(2.17) = 15.35$
We expect at least 75% of the cycles to fall in the interval 6.67 years to 15.35 years.

(b) $\bar{x} - 4s = 11.01 - 4(2.17) = 2.33$
$\bar{x} + 4s = 11.01 + 4(2.17) = 19.69$
We expect at least 93.8% of the cycles to fall in the interval 2.33 years to 19.69 years.

9. (a) $\text{Range} = 956 - 219 = 737$

$\bar{x} = \frac{\Sigma x}{n} = \frac{3968}{7} \approx 566.9$

(b) $s^2 = \frac{\Sigma(x - \bar{x})^2}{n-1} = \frac{427,213}{6} \approx 71,202$

$s = \sqrt{s^2} = \sqrt{71,202} \approx 266.8$

(c) $CV = \frac{s}{\overline{x}} \cdot 100 = \frac{266.8}{566.9} \cdot 100 = 47.1\%$

s is 47.1% of $\overline{x}$.

(d) $\overline{x} - 2s = 566.9 - 2(266.8) \approx 33$

$\overline{x} + 2s = 566.9 + 2(266.8) \approx 1100$

We expect at least 75% of the artifact counts for all such excavation sites to fall in the interval 33 to 1100.

11. $\overline{x} = 9.3$, $s = 2.6$ for domestic routine.

$\overline{x} - 2s = 9.3 - 2(2.6) = 4.1$

$\overline{x} + 2s = 9.3 + 2(2.6) = 14.5$

We would expect at least 75% of the percentages of domestic routine artifacts at different sites to fall in the interval 4.1% to 14.5%.

13. Construction artifacts have the largest mean and the smallest *CV*. Therefore, construction artifacts have the highest average and lowest relative standard deviation.

15. (a) Pax $CV = \frac{s}{\overline{x}} \cdot 100 = \frac{14.05}{9.58} \cdot 100 \approx 146.7\%$

Vanguard $CV = \frac{s}{\overline{x}} \cdot 100 = \frac{12.50}{9.02} \cdot 100 \approx 138.6\%$

Vanguard fund has slightly less risk for unit of return.

(b) Pax: $\overline{x} - 2s = 9.58 - 2(14.05) = -18.52$

$\overline{x} + 2s = 9.58 + 2(14.05) = 37.68$

At least 75% of the data fall in the interval −18.52% to 37.68%.

Vanguard: $\overline{x} - 2s = 9.02 - 2(12.50) = -15.98$

$\overline{x} + 2s = 9.02 + 2(12.50) = 34.02$

At least 75% of the data fall in the interval −15.98% to 34.02%.

Vanguard has a narrower range of returns, with less downside, but also less upside.

17. $CV = \frac{s}{\overline{x}} \cdot 100$

$\frac{\overline{x} \cdot CV}{100} = s$

$s = \frac{\overline{x} \cdot CV}{100}$

$s = \frac{2.2(1.5)}{100}$

$s = 0.033$

19. Answers vary.

Section 3.3

1.

Class	f	x	xf	$x-\bar{x}$	$(x-\bar{x})^2$	$(x-\bar{x})^2 f$
21–30	260	25.5	6630	−10.3	106.09	27,583.4
31–40	348	35.5	12,354	−0.3	0.09	31.3
41 and over	287	45.5	13,058.5	9.7	94.09	27,003.8
	$n=\Sigma f=895$		$\Sigma xf=32{,}042.5$			$\Sigma(x-\bar{x})^2 f=54{,}619$

$$\bar{x}=\frac{\Sigma xf}{n}=\frac{32{,}042.5}{895}\approx 35.80$$

$$s^2=\frac{\Sigma(x-\bar{x})^2\cdot f}{n-1}=\frac{54{,}619}{894}\approx 61.1$$

$$s=\sqrt{61.1}\approx 7.82$$

3.

Class	f	x	xf	$x-\bar{x}$	$(x-\bar{x})^2$	$(x-\bar{x})^2 f$
8.6–12.5	15	10.55	158.25	−5.05	25.502	382.537
12.6–16.5	20	14.55	291.00	−1.05	1.102	22.050
16.6–20.5	5	18.55	92.75	2.95	8.703	43.513
20.6–24.5	7	22.55	157.85	6.95	48.303	338.118
24.6–28.5	3	26.55	79.65	10.95	119.903	359.708
	$n=\Sigma f=50$		$\Sigma xf=779.5$			$\Sigma(x-\bar{x})^2 f=1145.9$

$$\bar{x}=\frac{\Sigma xf}{n}=\frac{779.5}{50}\approx 15.6$$

$$s^2=\frac{\Sigma(x-\bar{x})^2 f}{n-1}=\frac{1145.9}{49}\approx 23.4$$

$$s=\sqrt{23.4}\approx 4.8$$

5.

Class	f	x	xf	$x-\bar{x}$	$(x-\bar{x})^2$	$(x-\bar{x})^2 f$
18–24	78	21.0	1638.0	−18.12	328.33	25610.1
25–34	75	29.5	2212.5	−9.62	92.54	6940.8
35–44	48	39.5	1896.0	0.38	0.14	6.9
45–54	33	49.5	1633.5	10.38	107.74	3555.6
65–80	33	72.5	2392.5	33.38	1114.22	36769.4
	$n=\Sigma f=300$		$\Sigma xf=11{,}736$			$\Sigma(x-\bar{x})^2 f=86{,}589$

$$\bar{x} = \frac{\Sigma xf}{n} = \frac{11,736}{300} = 39.12$$

$$s = \sqrt{\frac{\Sigma(x-\bar{x})^2 f}{n-1}} = \sqrt{\frac{86,589}{299}} \approx 17.02$$

$$CV = \frac{s}{\bar{x}} \cdot 100 = \frac{17.02}{39.12} \cdot 100 \approx 43.5\%$$

7.

x	f	xf	$x^2 f$
3.5	2	7	24.5
4.5	2	9	40.5
5.5	4	22	121.0
6.5	22	143	929.5
7.5	64	480	3600.0
8.5	90	765	6502.5
9.5	14	133	1263.5
10.5	2	21	220.5
	$\Sigma f = 200$	$\Sigma xf = 1580$	$\Sigma x^2 f = 12,702$

$$\bar{x} = \frac{\Sigma xf}{n} = \frac{1580}{200} = 7.9$$

$$SS_x = \Sigma x^2 f - \frac{(\Sigma xf)^2}{n} = 12,702 - \frac{(1580)^2}{200} = 220$$

$$s = \sqrt{\frac{SS_x}{n-1}} = \sqrt{\frac{220}{199}} \approx 1.05$$

$$CV = \frac{s}{\bar{x}} \cdot 100 = \frac{1.05}{7.9} \cdot 100 \approx 13.29\%$$

9. (a)

x	f	xf	$x^2 f$
1	3	3	3
2	7	14	28
3	6	18	54
4	5	20	80
5	4	20	100
6	2	12	72
7	0	0	0
8	1	8	64
9	2	18	162
10	1	10	100
	$\Sigma f = 31$	$\Sigma xf = 123$	$\Sigma x^2 f = 663$

$$\bar{x} = \frac{\Sigma xf}{n} = \frac{123}{31} \approx 3.97$$

$$SS_x = \Sigma x^2 f - \frac{(\Sigma xf)^2}{n} = 663 - \frac{(123)^2}{31} \approx 175$$

$$s = \sqrt{\frac{SS_x}{n-1}} = \sqrt{\frac{175}{30}} \approx 2.415$$

(b) The results are the same.

11.

x	f	xf	$x^2 f$
2.8	145	406.0	1136.8
6.3	270	1701.0	10716.3
1.8	224	403.2	725.8
4.8	271	1300.8	6243.8
3.0	67	201.0	603.0
	$\Sigma f = 977$	$\Sigma xf = 4012$	$\Sigma x^2 f = 19{,}426$

$$\bar{x} = \frac{\Sigma xf}{n} = \frac{4012}{977} \approx 4.11$$

$$SS_x = \Sigma x^2 f - \frac{(\Sigma xf)^2}{n} = 19{,}426 - \frac{(4012)^2}{977} \approx 2951$$

$$s^2 = \frac{SS_x}{n-1} = \frac{2951}{976} \approx 3.02$$

$$s = \sqrt{3.02} \approx 1.74$$

Section 3.4

1. 82% or more of the scores were at or below her score. 100% – 82% = 18% or less of the scores were above her score. Note: This answer is correct, but it relies on a more precise definition than that given in the text on page 124. An adequate answer, matching the definition in the text would be: 82% of the scores were at or below her score, and (100 – 82)% = 18% of the scores were at or above her score.

3. No, the score 82 might have a percentile rank less than 70.

5. Order the data from smallest to largest.

Lowest value = 0.52
Highest value = 1.92

There are 16 data values.

$$\text{Median} = \frac{0.85+0.90}{2} = 0.875$$

There are 8 values less than 0.875 and 8 values greater than 0.875.

$$Q_1 = \frac{0.72+0.75}{2} = 0.735$$
$$Q_3 = \frac{1.15+1.50}{2} = 1.325$$
$$IQR = Q_3 - Q_1 = 1.325 - 0.735 = 0.59$$

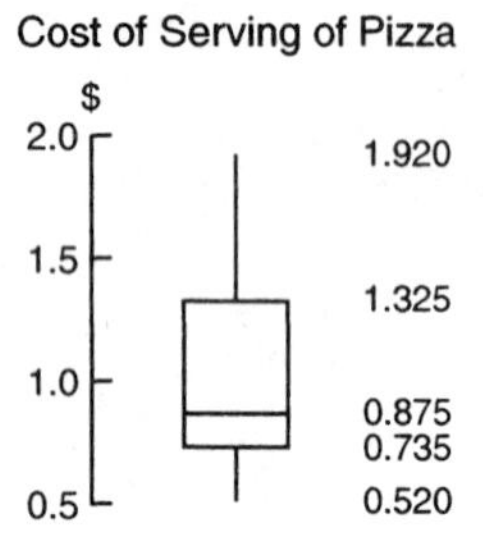

7. Order the data from smallest to largest.

Lowest value = 2
Highest value = 42

There are 20 data values.

$$\text{Median} = \frac{23+23}{2} = 23$$

There are 10 values less than the Q_2 position and 10 values greater than the Q_2 position.

$$Q_1 = \frac{8+11}{2} = 9.5$$
$$Q_3 = \frac{28+29}{2} = 28.5$$
$$IQR = Q_3 - Q_1 = 28.5 - 9.5 = 19$$

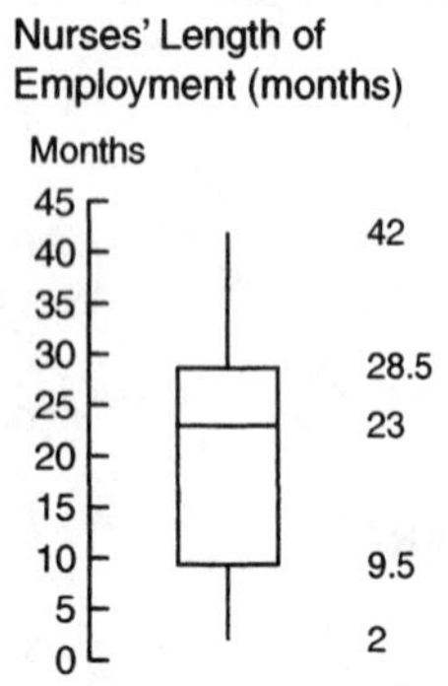

9. Order each set of data from smallest to largest.

Suburban

Lowest value = 808
Highest value = 1292

$$\text{Median} = \frac{992+1170}{2} = 1081$$

There are five values above and five values below the median.

$Q_1 = 972$
$Q_3 = 1216$
$IQR = 1216 - 972 = 244$

Urban

Lowest value = 1768
Highest value = 2910

$$\text{Median} = \frac{2107+2356}{2} = 2231.5$$

There are five values above and five values below the median.

$Q_1 = 1968$
$Q_3 = 2674$
$IQR = 2674 - 1968 = 706$

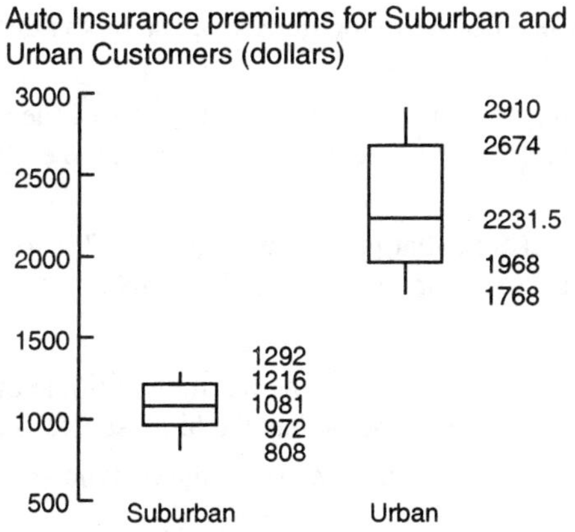

The entire box-and-whisker plot for urban is above that for suburban. Even the highest value for suburban is less than the lowest value for urban. The suburban data is less variable than that of urban data.

11. (a)

$$\text{Lowest value} = 17$$
$$\text{Highest value} = 38$$

There are 50 data values.

$$\text{Median} = \frac{24+24}{2} = 24$$

There are 25 values above and 25 values below the Q_2 position.

$$Q_1 = 22$$
$$Q_3 = 27$$
$$IQR = 27 - 22 = 5$$

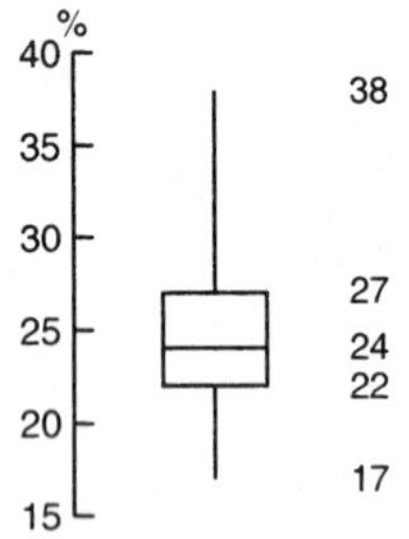

(b) 26% is in the 3rd quartile, since it is between the median and Q_3.

13. (a) California has the lowest premium since its left whisker is farthest to the left. Pennsylvania has the highest premium since its right whisker is farthest to the right.

(b) Pennsylvania has the highest median premium since its line in the middle of the box is farthest to the right.

(c) California has the smallest range of premiums since the distance between the ends of the whiskers is the smallest. Texas has the smallest interquartile range since the distance between the ends of the boxes is the smallest.

(d) Based on the answers to (a)-(c) above, we can determine that part (a) of Figure 3-13 is for Texas, part (b) of Figure 3-13 is for Pennsylvania, and part (c) of Figure 3-13 is for California.

15. (a) Assistant had the smallest median percentage salary increase since the bar in the middle of the box is the lowest. Associate had the single highest salary increase since it has the highest asterisk.

(b) Instructor had the largest spread between the first and third quartiles since the distance between the ends of the box is greatest.

(c) Assistant had the smallest spread for the lower 50% of the percentage salary increases since the distance between the bar in the box and the maximum value is the smallest.

(d) Professor had the most symmetric percentage salary increases because there are no outliers and the bar representing the median is close to the center of the box.
Yes, if the outliers for the associate professors were omitted, that distribution would appear to be symmetric.

(e) Associate professor:

$$IQR = 5.075 - 2.350 = 2.725$$
$$Q_3 + 1.5(IQR) = 5.075 + 1.5(2.725) \approx 9.16$$

Yes, since 17.7 is greater than 9.16, there is at least one outlier.

Instructor:

$$IQR = 5.800 - 2.850 = 2.950$$
$$Q_3 + 1.5(IQR) = 5.800 + 1.5(2.950) \approx 10.23$$

Yes, since 13.4 is greater than 10.23, there is at least one outlier.

Chapter 3 Review

1. (a) $\bar{x} = \frac{\Sigma x}{n} = \frac{876}{8} = 109.5$

$$s = \sqrt{\frac{\Sigma(x-\bar{x})^2}{n-1}} = \sqrt{\frac{7044}{7}} = \sqrt{1006.3} \approx 31.7$$

$$CV = \frac{s}{\bar{x}} \cdot 100 = \frac{31.7}{109.5} \cdot 100 \approx 28.9\%$$

range = maximum value − minimum value
$= 142 - 73 = 69$

(b) $\bar{x} = \frac{\Sigma x}{n} = \frac{881}{8} = 110.125$

$$s = \sqrt{\frac{\Sigma(x-\bar{x})^2}{n-1}} = \sqrt{\frac{358.87}{7}} \approx 7.2$$

$$CV = \frac{s}{\bar{x}} \cdot 100 = \frac{7.2}{110.125} \cdot 100 \approx 6.5\%$$

range = maximum value − minimum value
$= 120 - 100 = 20$

(c) The means are about the same. The first distribution has greater spread. The standard deviation, *CV*, and range for the first set of measurements are greater than those for the second set of measurements.

3. (a) Lowest value = 31
Highest value = 68

There are 60 data values.

$$\text{Median} = \frac{45+45}{2} = 45$$

There are 30 values above and 30 values below the Q_2 position.

$$Q_1 = \frac{40+40}{2} = 40$$
$$Q_3 = \frac{52+53}{2} = 52.5$$
$$IQR = 52.5 - 40 = 12.5$$

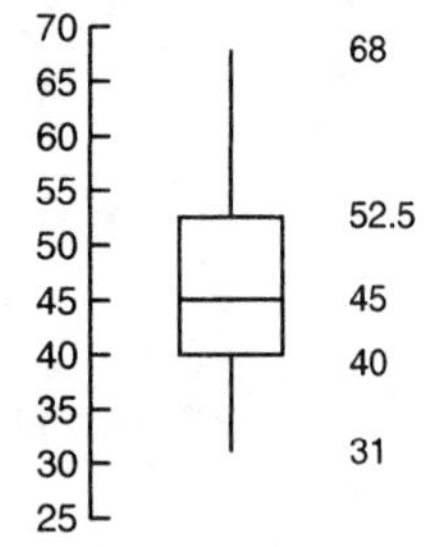

(b) Class width = 8

Class	x Midpoint	f	xf	x^2f
31–38	34.5	11	379.5	13,092.8
39–46	42.5	24	1020	43,350.0
47–54	50.5	15	757.5	38,253.8
55–62	58.5	7	409.5	23,955.8
63–70	66.5	3	199.5	13,266.8
		$n = \Sigma f = 60$	$\Sigma xf = 2766$	$\Sigma x^2 f = 131{,}919$

$$\bar{x} = \frac{\Sigma xf}{n} = \frac{2766}{60} = 46.1$$

$$s = \sqrt{\frac{\Sigma x^2 f - \frac{(\Sigma xf)^2}{n}}{n-1}} = \sqrt{\frac{131{,}919 - \frac{(2766)^2}{60}}{59}} = \sqrt{\frac{4406.4}{59}} \approx 8.64$$

$$\bar{x} - 2s = 46.1 - 2(8.64) = 28.82$$
$$\bar{x} + 2s = 46.1 + 2(8.64) = 63.38$$

We expect at least 75% of the data to fall in the interval 28.82 to 63.38.

(c) $\bar{x} = 46.15,\ s \approx 8.63$

5. Mean weight $= \dfrac{2500}{16} = 156.25$

The mean weight is 156.25 lb.

7. (a) Mean $= \bar{x} = \dfrac{\Sigma x}{n} = \dfrac{10.1 + 6.2 + \cdots + 5.7}{6} = \dfrac{47}{6} \approx 7.83$

$$s = \sqrt{\frac{\Sigma(x-\bar{x})^2}{n-1}} = \sqrt{\frac{26.913}{5}} \approx 2.32$$

$$CV = \frac{s}{\bar{x}} \cdot 100 = \frac{2.32}{7.83} \cdot 100 \approx 29.6\%$$

Range = largest value − smallest value
= 10.1 − 5.3 = 4.8

(b) Mean $= \bar{x} = \frac{\Sigma x}{n} = \frac{10.2+9.7+\cdots+10.1}{6} = \frac{59.7}{6} = 9.95$

$$s = \sqrt{\frac{\Sigma(x-\bar{x})^2}{n-1}} = \sqrt{\frac{0.415}{5}} \approx 0.29$$

$$CV = \frac{s}{\bar{x}} \cdot 100 = \frac{0.29}{9.95} \cdot 100 \approx 2.9\%$$

$$\begin{aligned}\text{Range} &= \text{largest value} - \text{smallest value}\\ &= 10.3 - 9.6 = 0.7\end{aligned}$$

(c) Second line has more consistent performance as reflected by the smaller standard deviation, *CV*, and range.

9. (a) No; 77% is an average, which is not a guarantee that it will be this value for each college graduate.

(b) $\bar{x} - 2s = 40,478 - 2(8,500) = 23,478$

$\bar{x} + 2s = 40,478 + 2(8,500) = 57,478$

We expect at least 75% of the earnings to fall in the interval \$23,478 to \$57,478.

(c) Weighted average $= \frac{\Sigma xw}{\Sigma w}$

$$= \frac{3500(0.51) + 6000(0.21) + 10,000(0.06) + 14,000(0.08) + 18,000(0.07) + 25,000(0.07)}{1}$$

$$= 7775$$

The weighted average is \$7,775.

11. Weighted average $= \frac{\Sigma xw}{\Sigma w}$

$$= \frac{5(2)+8(3)+7(3)+9(5)+7(3)}{2+3+3+5+3}$$

$$= \frac{121}{16}$$

$$\approx 7.56$$

13. (a) It is possible for the range and the standard deviation to be the same. For instance, for data values that are all the same, such as 1, 1, 1, 1, 1, the range and standard deviation are both 0.

(b) It is possible for the mean, median, and mode to be all the same. For instance, the data set 1, 2, 3, 3, 3, 4, 5 has mean, median, and mode all equal to 3. The three can also all be different, as in the data set 1, 2, 3, 3. In this case, the mean is 2.25, the median is 2.5, and the mode is 3.

Chapter 4 Elementary Probability Theory

Section 4.1

1. Answers vary. Probability is a number between 0 and 1, inclusive, that expresses the likelihood that a specific event will occur. Three ways to find or assign a probability to an event are (1) through intuition (subjective probability), (2) by considering the long-term relative frequency of recurrence of an event in repeated independent trials (empirical probability), and (3) by computing the ratio of the number of favorable outcomes to the total number of possible outcomes, assuming all outcomes are equally likely (classical probability).

3. These are not probabilities: (b) because it is greater than 1, (d) because it is less than zero (negative), (h) 150% = 1.50, because it is greater than 1.

5. Answers vary. The result is a sample, although not necessarily a good one, showing the relative frequency of people able to wiggle their ears.

7. **(a)** $P(\text{no similar preferences}) = P(0) = \frac{15}{375},\ P(1) = \frac{71}{375},\ P(2) = \frac{124}{375},\ P(3) = \frac{131}{375},\ P(4) = \frac{34}{375}$

(b) $\frac{15+71+124+131+34}{375} = \frac{375}{375} = 1$, yes

Personality types were classified into 4 main preferences; all possible numbers of shared preferences were considered. The sample space is 0, 1, 2, 3, and 4 shared preferences.

9. **(a)** Note: "includes the left limit but not the right limit" means 6 A.M. $\le$ time $t <$ noon, noon $\le t <$ 6 P.M., 6 P.M. $\le t <$ midnight, midnight $\le t <$ 6 A.M. $P(\text{best idea 6 A.M.–12 noon}) = \frac{290}{966} \approx 0.30$; $P(\text{best idea 12 noon–6 P.M.}) = \frac{135}{966} \approx 0.14$; $P(\text{best idea 6 P.M.–12 midnight}) \frac{319}{966} \approx 0.33$; $P(\text{best idea from 12 midnight to 6 A.M.}) = \frac{222}{966} \approx 0.23$.

(b) The probabilities add up to 1. They should add up to 1 provided that the intervals do not overlap and each inventor chose only one interval. The sample space is the set of four time intervals.

11. **(a)** Since the responses of all 1000 people surveyed have been accounted for (770 + 160 + 70 = 1000), the 3 responses describe the sample space. $P(\text{left alone}) = \frac{770}{1000} = 0.77$, $P(\text{waited on}) = \frac{160}{1000} = 0.16$, $P(\text{treated differently}) = \frac{70}{1000} = 0.07$

$0.77 + 0.16 + 0.07 = 1$

The probabilities do add to 1, which they should, because they are the sum of probabilities of all possible outcomes in the sample space.

(b) Complementary events: $P(\textit{not } A) = 1 - P(A)$:

$P(\text{do not want to be left alone}) = 1 - P(\text{left alone}) = 1 - 0.77 = 0.23$

$P(\text{do not want to be waited on}) = 1 - P(\text{waited on}) = 1 - 0.16 = 0.84$

13. **(a)** Given: odds in favor of A are $n{:}m$ $\left(\text{i.e., } \frac{n}{m}\right)$.

Show $P(A) = \frac{n}{m+n}$

Proof: odds in favor of A are $\frac{P(A)}{P(not\ A)}$ by definition

$$P(not\ A) = 1 - P(A) \quad \text{complementary events}$$

$$\frac{n}{m} = \frac{P(A)}{P(not\ A)} = \frac{P(A)}{1-P(A)} \quad \text{substitution}$$

$$n[1 - P(A)] = m[P(A)] \quad \text{cross multiply}$$

$$n - n[P(A)] = m[P(A)]$$

$$n = n[P(A)] + m[P(A)]$$

$$n = (n+m)[P(A)]$$

So $\frac{n}{n+m} = P(A)$ as was to be shown.

(b) Odds of a successful call = odds of sale are 2 to 15. 2 to 15 can be written as 2:15 or $\frac{2}{15}$ then from part (a): if the odds in favor of a sale are 2:15 (let $n = 2$, $m = 15$) then $P(\text{sale}) = \frac{n}{n+m} = \frac{2}{2+15}$

$= \frac{2}{17} \approx 0.118.$

(c) Odds of free throw are 3 to 5, i.e., 3:5
Let $n = 3$ and $m = 5$ here then from part (a):

$P(\text{free throw}) = \frac{n}{n+m} = \frac{3}{3+5} = \frac{3}{8} = 0.375.$

15. Make a table showing the information known about the 127 people who walked by the store: [Example 6 in Section 4.2 uses this technique.]

	Buy	Did not buy	Row Total
Came into the store	25	58 – 25 = 33	58
Did not come in	0	69	127 – 58 = 69
Column Total	25	102	127

If 58 came in, 69 didn't; 25 of the 58 bought something, so 33 came in but didn't buy anything. Those who did not come in, couldn't buy anything. The row entries must sum to the row totals; the column entries must sum to the column totals; and the row totals, as well as the column totals, must sum to the overall total, i.e., the 127 people who walked by the store. Also, the four inner cells must sum to the overall total: 25 + 33 + 0 + 69 = 127.

This kind of problem relies on formula (2), $P(\text{event } A) = \frac{\text{number outcomes favorable to } A}{\text{total number of outcomes}}$. The "trick" is to decide what belongs in the denominator *first*. If the denominator is a row total, stay in that row. If the denominator is a column total, stay in that column. If the denominator is the overall total, the numerator can be a row total, a column total, or the number in any one of the four "cells" inside the table.

(a) total outcomes: people walking by, overall total, 127
favorable outcomes: enter the store, row total, 58 (that's all we know about them)

$P(A) = \frac{58}{127} \approx 0.46$

(b) total outcomes: people who walk into the store, row total 58
favorable outcomes: staying in the row, those who buy: 25

$$P(A)=\frac{25}{58}\approx 0.43$$

(c) total outcomes: people walking by, overall total 127
favorable outcomes: people coming in *and* buying, the cell at the *intersection* of the "coming in" row and the "buying" column (the upper left corner), 25 (Recall from set theory that "and" means both things happen, that the two sets *intersect*: >)

$$P(A)=\frac{25}{127}\approx 0.20$$

(d) total outcomes: people coming into the store, row total, 58
favorable outcomes: staying in the row, those who do not buy, 33

$$P(A)=\frac{33}{58}\approx 0.57$$

$$\left(\text{alternate method: this is the complement to (b): } P(A)=1-\frac{25}{58}=\frac{33}{58}\approx 0.57\right)$$

Section 4.2

1. (a) Green and blue are mutually exclusive because each M&M candy is only 1 color.
$P(\text{green } \textit{or} \text{ blue}) = P(\text{green}) + P(\text{blue}) = 10\% + 10\% = 20\% = 0.20$

(b) Yellow and red are mutually exclusive, again, because each candy is only one color, and if the candy is yellow, it can't be red, too.
$P(\text{yellow } \textit{or} \text{ red}) = P(\text{yellow}) + P(\text{red}) = 20\% + 20\% = 40\% = 0.40$

(c) It is faster here to use the complementary event rule than to add up the probabilities of all the colors except purple.
$P(\textit{not} \text{ purple}) = 1 - P(\text{purple}) = 1 - 0.20 = 0.80$, or 80%

3. (a) Mutually exclusive: $P(\text{green } \textit{or} \text{ blue}) = P(\text{green}) + P(\text{blue}) = 16.6\% + 16.6\% = 33.2\% = 0.332$

(b) Mutually exclusive: $P(\text{yellow } \textit{or} \text{ red}) = P(\text{yellow}) + P(\text{red}) = 16.6\% + 16.6\% = 33.2\% = 0.332$

(c) $P(\textit{not} \text{ purple}) = 1 - P(\text{purple}) = 1 - 0 = 1$ (there is no purple)
Since the color distributions differ for plain and caramel M&Ms, if the answers were the same, it would only be by coincidence.

Hint for Problems 5–7: Refer to Figure 4–2 if necessary. (Without loss of generality, let the red die be the first die and the green die be the second die in Figure 4–2.) Think of the outcomes as an (x, y) ordered pair. Then, without loss of generality, (1, 6) means 1 on the red die and 6 on the green die. (We are "ordering" the dice for convenience only – which is first and which is second have no bearing on this problem.) The only important fact is that they are distinguishable outcomes, so that (1 on red, 2 on green) is different from (2 on red, 1 on green).

5. (a) Yes, the outcome of the red die does not influence the outcome of the green die.

(b) $P(5 \text{ on green } \textit{and} \text{ 3 on red}) = P(5 \text{ on green}) \cdot P(3 \text{ on red}) = \left(\frac{1}{6}\right)\left(\frac{1}{6}\right)=\frac{1}{36}\approx 0.028$ because they are independent.

(c) $P(3 \text{ on green } \textit{and} \text{ 5 on red}) = P(3 \text{ on green}) \cdot P(5 \text{ on red}) = \left(\frac{1}{6}\right)\left(\frac{1}{6}\right)=\frac{1}{36}\approx 0.028$

(d) P[(5 on green *and* 3 on red) *or* (3 on green *and* 5 on red)]
$= P$(5 on green *and* 3 on red) + P(3 on green *and* 5 on red)]
$= \frac{1}{36} + \frac{1}{36} = \frac{2}{36} = \frac{1}{18} \approx 0.056$ [because they are mutually exclusive outcomes]

7. (a) $1 + 5 = 6, 2 + 4 = 6, 3 + 3 = 6, 4 + 2 = 6, 5 + 1 = 6$
$P(\text{sum} = 6) = P$[(1, 5) *or* (2, 4) *or* (3 on red, 3 on green) *or* (4, 2) *or* (5, 1)]
$= P(1, 5) + P(2, 4) + P(3, 3) + P(4, 2) + P(5, 1)$
since the (red, green) outcomes are mutually exclusive
$= \left(\frac{1}{6}\right)\left(\frac{1}{6}\right) + \left(\frac{1}{6}\right)\left(\frac{1}{6}\right) + \left(\frac{1}{6}\right)\left(\frac{1}{6}\right) + \left(\frac{1}{6}\right)\left(\frac{1}{6}\right) + \left(\frac{1}{6}\right)\left(\frac{1}{6}\right)$
because the red die outcome is independent of the green die outcome
$= \frac{1}{36} + \frac{1}{36} + \frac{1}{36} + \frac{1}{36} + \frac{1}{36} = \frac{5}{36}$

(b) $1 + 3 = 4, 2 + 2 = 4, 3 + 1 = 4$
$P(\text{sum is } 4) = P$[(1, 3) *or* (2, 2) *or* (3, 1)]
$= P(1, 3) + P(2, 2) + P(3, 1)$
because the (red, green) outcomes are mutually exclusive
$= \left(\frac{1}{6}\right)\left(\frac{1}{6}\right) + \left(\frac{1}{6}\right)\left(\frac{1}{6}\right) + \left(\frac{1}{6}\right)\left(\frac{1}{6}\right)$
because the red die outcome is independent of the green die outcome
$= \frac{1}{36} + \frac{1}{36} + \frac{1}{36} = \frac{3}{36} = \frac{1}{12}$

(c) Since a sum of six can't simultaneously be a sum of four, these are mutually exclusive events;
P(sum of 6 *or* 4) = P(sum of 6) + P(sum of 4) = $\frac{5}{36} + \frac{3}{36} = \frac{8}{36} = \frac{2}{9}$

9. (a) No, the key idea here is "without replacement," which means the draws are dependent, because the outcome of the second card drawn depends on what the first card drawn was. Let the card draws be represented by an (x, y) ordered pair. For example, (K, 6) means the first card drawn was a king and the second card drawn was a 6. Here the order of the cards is important.

(b) P(ace on 1st *and* king on second) = P(ace, king) $= \left(\frac{4}{52}\right)\left(\frac{4}{51}\right) = \frac{16}{2652} = \frac{4}{663}$

There are 4 aces and 4 kings in the deck. Once the first card is drawn and not replaced, there are only 51 cards left to draw from, but all the kings are still there.

(c) P(king, ace) $= \left(\frac{4}{52}\right)\left(\frac{4}{51}\right) = \frac{16}{2652} = \frac{4}{663}$

There are 4 kings and 4 aces in the deck. Once the first card is drawn and not replaced, there are only 51 cards left to draw from, but all the aces are still there.

(d) P(ace *and* king in either order)
$= P$[(ace, king) *or* (king, ace)]
$= P$(ace, king) + P(king, ace) because these two outcomes are mutually exclusive
$= \frac{16}{2652} + \frac{16}{2652} = \frac{32}{2652} = \frac{8}{663}$

11. (a) Yes; the key idea here is "with replacement." When the first card drawn is replaced, the sample space is the same when the second card is drawn as it was when the first card was drawn and the second card is in no way influenced by the outcome of the first draw; in fact, it is possible to draw the same card twice. Let the card draws be represented by an (x, y) ordered pair; for example (K, 6) means a king was drawn first, replaced, and then the second card, a "6," was drawn independently of the first.

(b) $P(\text{A, K}) = P(\text{A}) \cdot P(\text{K})$ because they are independent

$$= \left(\frac{4}{52}\right)\left(\frac{4}{52}\right) = \frac{16}{2704} = \frac{1}{169}$$

(c) $P(\text{K, A}) = P(\text{K}) \cdot P(\text{A})$ because they are independent

$$= \left(\frac{4}{52}\right)\left(\frac{4}{52}\right) = \frac{16}{2704} = \frac{1}{169}$$

(d) $P[(\text{A, K})\ or\ (\text{K, A})] = P(\text{A, K}) + P(\text{K, A})$ since the 2 outcomes are mutually exclusive when we consider the order

$$= \frac{1}{169} + \frac{1}{169} = \frac{2}{169}$$

13. (a) $P(6\ or\ \text{older}) = P[(6\text{ to }9)\ or\ (10\text{ to }12)\ or\ (13\text{ and over})]$

$= P(6-9) + P(10-12) + P(13+)$ because they are mutually exclusive age groups-no child is both 7 and 11 years old.

$= 27\% + 14\% + 22\% = 63\% = 0.63$

(b) $P(12\text{ or younger}) = 1 - P(13\text{ and over}) = 1 - 0.22 = 0.78$

(c) $P(\text{between 6 and 12}) = P[(6\text{ to }9)\ or\ (10\text{ to }12)]$

$= P(6\text{ to }9) + P(10\text{ to }12)$ because the age groups are mutually exclusive

$= 27\% + 14\% = 41\% = 0.41$

(d) $P(\text{between 3 and 9}) = P[(3\text{ to }5)\ or\ (6\text{ to }9)]$

$= P(3\text{ to }5) + P(6\text{ to }9)$ because age categories are mutually exclusive

$= 22\% + 27\% = 49\% = 0.49$

Answers vary; however, category 10–12 years covers only 3 years while 13 and over covers many more years and many more people, including adults who buy toys for themselves.

15. What we know: P(polygraph says "lying" when person is lying) = 72%

P(polygraph says "lying" when person is not lying) = 7%

Let L denote that the polygraph results show lying and *not* L denote that the polygraph results show the person is not lying. Let T denote that the person is telling the truth and let *not* T denote that the person is not telling the truth, so $P(\text{L}, \textit{given not}\ \text{T}) = 72\%$

$P(\text{L}, \textit{given}\ \text{T}) = 7\%$.

We are told whether the person is telling the truth or not; what we know is what the polygraph results are, given the case where the person tells the truth, and given the situation where the person is not telling the truth.

(a) $P(\text{T}) = 0.90$ so $P(not\ \text{T}) = 0.10$

P(polygraph says lying and person tells truth)

$= \text{P(L } and \text{ T)} = P(\text{T}) \cdot P(\text{L}, \textit{given}\ \text{T})$

$= (0.90)(0.07) = 0.063 = 6.3\%$

(b) $P(not\ \text{T}) = 0.10$ so $P(\text{T}) = 0.90$

P(polygraph says lying and person is not telling the truth)

$= \text{P(L } and\ not \text{ T)} = P(not\ \text{T}) \cdot P(\text{L}, \textit{given not}\ \text{T})$

$= (0.10)(0.72) = 0.072 = 7.2\%$

(c) $P(\text{T}) = P(not\ \text{T}) = 0.50$

(a) $\text{P(L } and \text{ T)} = P(\text{T}) \cdot P(\text{L}, \textit{given}\ \text{T})$

$= (0.50)(0.07) = 0.035 = 3.5\%$

(b) $\text{P(L } and\ not \text{ T)} = P(not\ \text{T}) \cdot P(\text{L}, \textit{given not}\ \text{T})$

$= (0.50)(0.72) = 0.36 = 36\%$

(d) $P(\text{T}) = 0.15$ so $P(not\ \text{T}) = 1 - P(\text{T}) = 1 - 0.15 = 0.85$

(a) $P(\text{L}\ and\ \text{T}) = P(\text{T}) \cdot P(\text{L},\ given\ \text{T})$
$= (0.15)(0.07) = 0.0105 = 1.05\%$

(b) $P(\text{L}\ and\ not\ \text{T}) = P(not\ \text{T}) \cdot P(\text{L},\ given\ not\ \text{T})$
$= (0.85)(0.72) = 0.612 = 61.2\%$

17. Let E denote eyeglasses, C denote contact lenses, W denote women, and M denote men. Then we have $P(\text{E}) = 56\%$, $P(\text{C}) = 3.6\%$, $P(\text{W},\ given\ \text{E}) = 55.4\%$, $P(\text{M},\ given\ \text{E}) = 44.6\%$, $P(\text{W},\ given\ \text{C}) = 63.1\%$, $P(\text{M},\ given\ \text{C}) = 36.9\%$, and $P(\text{E}\ and\ \text{C}) = 0$.

(a) $P(\text{W}\ and\ \text{E}) = P(\text{E}) \cdot P(\text{W},\ given\ \text{E})$
using a conditional probability rule
$= (0.56)(0.554) \approx 0.310$

(b) $P(\text{M}\ and\ \text{E}) = P(\text{E}) \cdot P(\text{M},\ given\ \text{E})$
$= (0.56)(0.446) \approx 0.250$

(c) $P(\text{W}\ and\ \text{C}) = P(\text{C}) \cdot P(\text{W},\ given\ \text{C})$
$= (0.036)(0.631) \approx 0.023$

(d) $P(\text{M}\ and\ \text{C}) = P(\text{C}) \cdot P(\text{M},\ given\ \text{C})$
$= (0.036)(0.369) \approx 0.013$

(e) $P(\text{none of the above}) = 1 - [P(\text{W}\ and\ \text{E}) + P(\text{M}\ and\ \text{E}) + P(\text{W}\ and\ \text{C}) + P(\text{M}\ and\ \text{C})]$
$= 1 - [0.310 + 0.250 + 0.023 + 0.013]$
$= 1 - 0.596 = 0.404$

19. We have $P(A) = \frac{580}{1160}$, $P(Pa) = \frac{580}{1160} = P(not\ A)$, $P(S) = \frac{686}{1160}$, $P(N) = \frac{474}{1160} = P(not\ S)$

(a) $P(S) = \frac{686}{1160}$

$P(S,\ given\ A) = \frac{270}{580}$ (*given A* means stay in the *A*, aggressive row)

$P(S,\ given\ Pa) = \frac{416}{580}$ (staying in row *Pa*)

(b) $P(S) = \frac{686}{1160} = \frac{343}{580}$

$P(S,\ given\ Pa) = \frac{416}{580}$

They are not independent since the probabilities are not the same.

(c) $P(A\ and\ S) = P(A) \cdot P(S,\ given\ A)$
$= \left(\frac{580}{1160}\right)\left(\frac{270}{580}\right) = \frac{270}{1160}$

$P(Pa\ and\ S) = P(Pa) \cdot P(S,\ given\ Pa)$
$= \left(\frac{580}{1160}\right)\left(\frac{416}{580}\right) = \frac{416}{1160}$

(d) $P(N) = \frac{474}{1160}$

$P(N,\ given\ A) = \frac{310}{580}$ (stay in the *A* row)

(e) $P(N) = \frac{474}{1160} = \frac{237}{580}$

$P(N, \textit{given } A) = \frac{310}{580}$

Since the probabilities are not the same, *N* and *A* are not independent.

(f) $P(A \text{ or } S) = P(A) + P(S) - P(A \text{ and } S)$

$= \frac{580}{1160} + \frac{686}{1160} - \frac{270}{1160} = \frac{996}{1160}$

21. Let C denote the condition is present, and *not* C denote the condition is absent.

(a) $P(+, \textit{given } \text{C}) = \frac{72}{154}$ (stay in C column)

(b) $P(-, \textit{given } \text{C}) = \frac{82}{154}$ (stay in C column)

(c) $P(-, \textit{given not } \text{C}) = \frac{79}{116}$ (stay in *not* C column)

(d) $P(+, \textit{given not } \text{C}) = \frac{37}{116}$ (stay in *not* C column)

(e) $P(\text{C } \textit{and } +) = P(\text{C}) \cdot P(+, \textit{given } \text{C}) = \left(\frac{154}{270}\right)\left(\frac{72}{154}\right) = \frac{72}{270}$

(f) $P(\text{C } \textit{and } -) = P(\text{C}) \cdot P(-, \textit{given } \text{C}) = \left(\frac{154}{270}\right)\left(\frac{82}{154}\right) = \frac{82}{270}$

23. First determine the denominator. If it is a row or column total, the numerator will be in the body (inside) of the table in that same row or column. If the denominator is the grand total, the numerator can be one or more row totals, one or more column totals, or a body-of-the-table cell entry. A cell entry is usually indicated when the problem mentions both a row category and a column category, in which case the desired cell is the one where the row and column intersect.

(a) $P(2+) = 1 - P(< 2) = 1 - P(1 \text{ visit}) = 1 - \frac{962}{1894} = \frac{932}{1894}$

There is no additional information about the customer, so the denominator is the grand total.

(b) $P(2+, \textit{given } 25\text{–}39 \text{ years old}) = 1 - P(< 2, \textit{given } 25\text{–}39) = 1 - \frac{386}{739} = \frac{353}{739}$

The customer's age is given, so the denominator is the 25–39 row total, and the numerator is determined by the numbers in that row, using the complimentary rule for probabilities.

(c) $P(> 3 \text{ visits}) = P(4 \text{ or } 5 \text{ or } 6 \text{ or more visits})$

$= P(4) + P(5) + P(6+)$

$= \frac{66}{1894} + \frac{44}{1894} + \frac{32}{1894} = \frac{142}{1894}$

There is no qualifier on the customer, so we use the grand total, 1894, as the denominator.

(d) $P(>3, \textit{given} \text{ age } 65+) = 1 - P(\leq 3, \textit{given } 65+)$
$= 1 - P[(1, 2, \textit{or } 3 \text{ visits}), \textit{given } 65+]$
$= 1 - [P(1 \text{ visit}, \textit{given } 65+) + P(2 \text{ visits}, \textit{given } 65+) + P(3 \text{ visits}, \textit{given } 65+)]$
$$= 1 - \left(\frac{115}{224} + \frac{69}{224} + \frac{18}{224}\right) = 1 - \frac{202}{224} = \frac{22}{224}$$

Alternately, this can be solved without the complementary event rule:
$P(>3 \text{ visits}, \textit{given } 65 \text{ or older}) = P(4 \textit{ or } 5 \textit{ or } 6 \text{ or more visits}, \textit{given } 65+)$
$= P(4, \textit{given } 65+) + P(5, \textit{given } 65+) + P(6 \text{ or more}, \textit{given } 65+)$
$$= \frac{12}{224} + \frac{7}{224} + \frac{3}{224} = \frac{22}{224}$$

Here, the complementary event rule has no advantage, work-wise, over the regular method.

(e) $P(40 \text{ or older}) = P(40\text{–}49 \textit{ or } 50\text{–}64 \textit{ or } 65+)$
$= P(40\text{–}49) + P(50\text{–}64) + P(65+)$
$$= \frac{434}{1894} + \frac{349}{1894} + \frac{224}{1894} = \frac{1007}{1894}$$

(f) $P(40 \text{ or older}, \textit{given } 4 \text{ visits}) = P(40\text{–}49 \textit{ or } 50\text{–}64 \textit{ or } 65+ \text{or over}, \textit{given } 4 \text{ visits})$
$= P(40\text{–}49, \textit{given } 4) + P(50\text{–}64, \textit{given } 4) + P(65+, \textit{given } 4)$
$$= \frac{13}{66} + \frac{14}{66} + \frac{12}{66} = \frac{39}{66}$$

The given 4 visits means the denominator is the 4 visit column total, and the numerator is composed from numbers in that column.

(g) $P(25\text{–}39 \text{ years old}) = \frac{739}{1894} \approx 0.3902$

$P(\text{visits more than once a week}) = 1 - P(\text{visits once}) = 1 - \frac{962}{1894} = \frac{932}{1894} \approx 0.4921$

$P(\text{visits} > 1, \textit{given } 25\text{–}39) = 1 - P(1 \text{ visit}, \textit{given } 25\text{–}39) = 1 - \frac{386}{739} = \frac{353}{739} \approx 0.4777$

25–39 years old is independent of more than 1 visit if $P(> 1 \text{ visit}) = P(> 1 \text{ visit}, \textit{given} \text{ age } 25\text{–}39)$ but $0.4921 \neq 0.4777$ so they are not independent.

25. Given: Let A be the event that a new store grosses > \$940,000 in year 1; then not A is the event the new store grosses ≤ \$940,000 the first year.
Let B be the event that the store grosses > \$940,000 in the second year; then not B is the event the store grosses ≤ \$940,000 in the second year of operation.

2 year results	Translations
A and B	profitable both years
A and *not* B	profitable first but not second year
not A and B	profitable second but not first year
not A and *not* B	not profitable either year

$P(A) = 65\%$ (show profit in first year)
$P(\textit{not } A) = 35\%$
$P(B) = 71\%$ (show profit in second year)
$P(\textit{not } B) = 29\%$
$P(\text{close}) = P(\textit{not } A \text{ and } \textit{not } B)$
$P(B, \textit{given } A) = 87\%$

(a) $P(A) = 65\% = 0.65$ (from the given)

(b) $P(B) = 71\% = 0.71$ (from the given)

(c) $P(B, \textit{given } A) = 87\% = 0.87$ (from the given)

(d) $P(A \text{ and } B) = P(A) \cdot P(B, \text{given } A) = (0.65)(0.87) = 0.5655 \approx 0.57$ (conditional probability rule)

(e) $P(A \text{ or } B) = P(A) + P(B) - P(A \text{ and } B) = 0.65 + 0.71 - 0.57 = 0.79$ (addition rule)

(f) $P(\text{not closed}) = P(\text{show a profit in year 1 or year 2 or both}) = 0.79$ (same question as (e))
$P(\text{closed}) = 1 - P(\text{not closed}) = 1 - 0.79 = 0.21$ (complimentary event rule)

27. Let TB denote that the person has tuberculosis, so *not* TB denotes that the person does not have tuberculosis.
Let + indicate the test for tuberculosis indicates the presence of the disease, so – indicates that the test for tuberculosis shows no disease.
Given: $P(+, \textit{given } \text{TB}) = 0.82$ (sensitivity of the test)
$P(+, \textit{given not } \text{TB}) = 0.09$ (false-positive rate)
$P(\text{TB}) = 0.04$

(a) $P(\text{TB } \textit{and } +) = P(\text{TB}) \cdot P(+, \textit{ given } \text{TB})$ (by the conditional probability rule)
$= (0.04)(0.82) = 0.0328 \approx 0.033$ (predictive value of the test)

(b) $P(\textit{not } \text{TB}) = 1 - P(\text{TB})$ (complementary events)
$= 1 - 0.04 = 0.96$

(c) $P(\textit{not } \text{TB } \textit{and } +) = P(\textit{not } \text{TB}) \cdot P(+, \textit{given not } \text{TB})$ (conditional probability rule and (b))
$= (0.96)(0.09) = 0.0864 \approx 0.086$

(Note: refer to #20 above to see terminology.)

Section 4.3

1. (a) Outcomes for Tossing a Coin Three Times

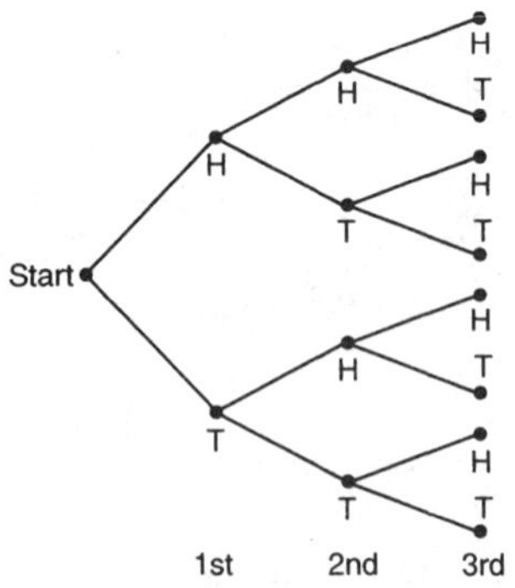

(b) HHT, HTH, THH: 3

(c) 8 possible outcomes, 3 with exactly 2 Hs: $\frac{3}{8}$

3. (a) Outcomes for Drawing Two Balls (without replacement)

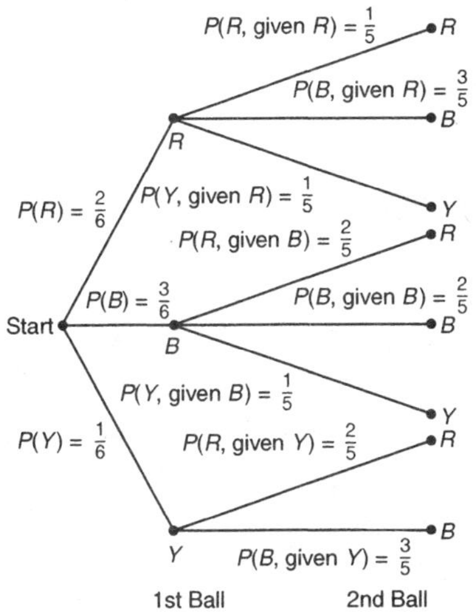

Because we drew without replacement the number of available balls drops to 5 and one of the colors drops by 1. Note that if the yellow ball is drawn first, there are only two possibilities for the second draw: red and blue; the yellow balls are exhausted.

(b) $P(R, R) = \left(\frac{2}{6}\right)\left(\frac{1}{5}\right) = \frac{2}{30} = \frac{1}{15}$

$P(R, B) = \left(\frac{2}{6}\right)\left(\frac{3}{5}\right) = \frac{6}{30} = \frac{1}{5}$

$P(R, Y) = \left(\frac{2}{6}\right)\left(\frac{1}{5}\right) = \frac{2}{30} = \frac{1}{15}$

$P(B, R) = \left(\frac{3}{6}\right)\left(\frac{2}{5}\right) = \frac{6}{30} = \frac{1}{5}$

$P(B, B) = \left(\frac{3}{6}\right)\left(\frac{2}{5}\right) = \frac{6}{30} = \frac{1}{5}$

$P(B, Y) = \left(\frac{3}{6}\right)\left(\frac{1}{5}\right) = \frac{3}{30} = \frac{1}{10}$

$P(Y, R) = \left(\frac{1}{6}\right)\left(\frac{2}{5}\right) = \frac{2}{30} = \frac{1}{15}$

$P(Y, B) = \left(\frac{1}{6}\right)\left(\frac{3}{5}\right) = \frac{3}{30} = \frac{1}{10}$

where $P(x, y)$ is the probability the first ball is color x, and the second ball is color y. Multiply the branch probability values along each branch from start to finish. Observe the sum of the probabilities is 1.

5. (a) Choices for Three True/False Questions

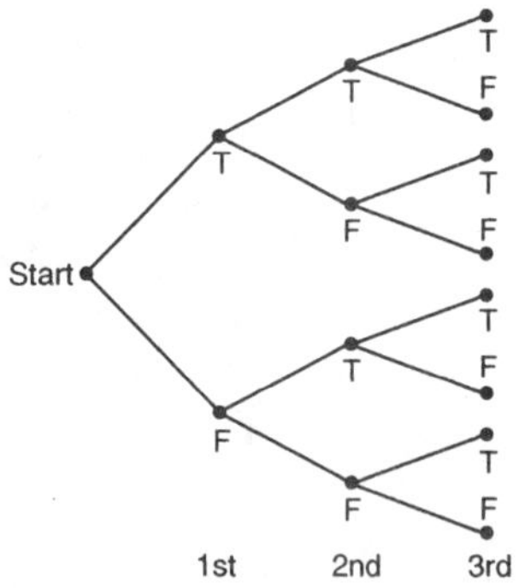

The tree diagram looks exactly like that of problem 1, because each event has 2 outcomes, and the events are independent, so possible outcomes for the second event are the same as for the first event.

(b) $P(\text{3 correct responses}) = \left(\frac{1}{2}\right)\left(\frac{1}{2}\right)\left(\frac{1}{2}\right) = \left(\frac{1}{2}\right)^3 = \frac{1}{8}$

7. 4 wire choices for the first leaves 3 wire choices for the second, 2 for the third, and only 1 wire choice for the fourth wire connection: $4 \cdot 3 \cdot 2 \cdot 1 = 4! = 24$.

9. (a) Choose 1 card from each deck. The number of pairs (one card from the first deck and one card from the second) is $52 \cdot 52 = 52^2 = 2704$.

(b) There are 4 kings in the first deck and four in the second, so $4 \cdot 4 = 16$.

(c) There are 16 ways to draw a king from each deck, and 2704 ways to draw a card from each deck, so $\frac{16}{2704} = \frac{1}{169} \approx 0.006$.

11. There are 4 fertilizers to choose from, 3 temperature zones to choose from for each fertilizer, and 3 possible water treatments for every fertilizer-temperature zone combination: $4 \cdot 3 \cdot 3 = 36$.

Problems 13, and 15 deal with permutations, $P_{n,r} = \frac{n!}{(n-r)!}$. This counts the number of ways r objects can be selected from n when the order of the result is important. For example, if we choose two people from a group, the first of which is to be the group's chair, and the second, the assistant chair, then (John, Mary) is distinct from (Mary, John).

13. $P_{5,2} : n = 5,\ r = 2$

$$P_{5,2} = \frac{5!}{(5-2)!} = \frac{5 \cdot 4 \cdot 3 \cdot 2 \cdot 1}{3!} = 20$$

15. $P_{7,7} : n = r = 7$

$$P_{7,7} = \frac{7!}{(7-7)!} = \frac{7!}{0!} = 7! = 5040 \text{ (recall } 0! = 1)$$

In general, $P_{n,n} = \frac{n!}{(n-n)!} = \frac{n!}{0!} = \frac{n!}{1} = n!$.

Problems 17, and 19 deal with combination, $C_{n,r} = \dfrac{n!}{r!(n-r)!}$. This counts the number of ways r items can be selected from among n items when the order of the result doesn't matter. For example, when choosing two people from an office to pick up coffee and doughnuts, (John, Mary) is the same as (Mary, John) – both get to carry the goodies back to the office.

17. $C_{5,2} : n = 5,\ r = 2$

$$C_{5,2} = \frac{5!}{2!(5-2)!} = \frac{5!}{2!3!} = \frac{5\cdot4\cdot3\cdot2\cdot1}{2\cdot1\cdot3\cdot2\cdot1} = \frac{20}{2} = 10$$

19. $C_{7,7} : n = r = 7$

$$C_{7,7} = \frac{7!}{7!(7-7)!} = \frac{7!}{7!0!} = \frac{7!}{7!(1)} = 1 \text{ (recall } 0! = 1)$$

In general, $C_{n,n} = \dfrac{n!}{n!(n-n)!} = \dfrac{n!}{n!0!} = \dfrac{n!}{n!(1)} = 1$. There is only 1 way to choose all n objects without regard to order.

21. Since the order matters (first is day supervisor, second is night supervisor, and third is coordinator), this is a permutation of 15 nurse candidates to fill 3 positions.

$$P_{15,3} = \frac{15!}{(15-3)!} = \frac{15!}{12!} = \frac{15\cdot14\cdot13\cdot12!}{12!} = 2730$$

23. **(a)** Think of this as 2 urns, the first having 8 red balls numbered 1 to 8, indicating the word to be defined, and the second urn containing 8 blue balls numbered 1 to 8, representing the possible definitions. Draw 1 ball from each urn without replacement. The (red number, blue number) pair indicates which word is "matched" with which definition–maybe correctly matched, maybe not. Here, (red 4, blue 6) is different from (red 6, blue 4)

$n = r = 8$

$$P_{8,8} = \frac{8!}{(8-8)!} = \frac{8!}{0!} = \frac{8!}{1} = 40{,}320$$

(b) This is the same as part (a), except we throw 3 definitions out; 3 words will be left with no definition.

$$P_{8,5} = \frac{8!}{(8-5)!} = \frac{8!}{3!} = \frac{40{,}320}{6} = 6720$$

25. Order matters because the resulting sequence determines who wins first place, who wins second place, and who wins third place.

$$P_{5,3} = \frac{5!}{(5-3)!} = \frac{5!}{2!} = \frac{120}{2} = 60$$

27. The order of trainee selection doesn't matter, since they are all going to be trained the same.

$$C_{15,5} = \frac{15!}{5!(15-5)!} = \frac{15!}{5!10!} = \frac{15\cdot14\cdot13\cdot12\cdot11\cdot10!}{5!10!} = \frac{15\cdot14\cdot13\cdot12\cdot11}{5\cdot4\cdot3\cdot2\cdot1} = 3003$$

29. **(a)** Six applicants are selected from twelve without regard to order.

$$C_{12,6} = \frac{12!}{6!6!} = \frac{479{,}001{,}600}{(720)^2} = 924$$

(b) There are 7 women and 5 men. This problem really asks, in how many ways can 6 women be selected from among 7, and zero men be selected from 5?

$$(C_{7,6})(C_{5,0}) = \left(\frac{7!}{6!(7-6)!}\right)\left(\frac{5!}{0!(5-0)!}\right) = \frac{7!}{6!1!}\cdot\frac{5!}{0!5!} = 7$$

Since the zero men are "selected" be default, all positions being filled. This problem reduces to, in how many ways can 6 applicants be selected from 7 women?

(c) $P(\text{event A}) = \dfrac{\text{number of favorable outcomes}}{\text{total number of outcomes}}$

$$P(\text{all hired are women}) = \frac{7}{924} = \frac{1}{132} \approx 0.008$$

Chapter 4 Review

1. $P(\text{asked}) = 24\% = 0.24$
$P(\text{received, } given \text{ asked}) = 45\% = 0.45$
$P(\text{asked } and \text{ received}) = P(\text{asked}) \cdot P(\text{received, } given \text{ asked}) = (0.24)(0.45) = 0.108 = 10.8\%$

3. (a) If the first card is replaced before the second is chosen (sampling with replacement), they are independent. If the sampling is without replacements they are dependent.

(b) $P(\text{heart}) = \dfrac{13}{52} = \dfrac{1}{4}$

with replacement, independent

$$P(\text{H on both}) = \left(\frac{1}{4}\right)\left(\frac{1}{4}\right) = \frac{1}{16} = 0.0625 \approx 0.063$$

(c) without replacement, dependent

$$P(\text{H on first } and \text{ H on second}) = \frac{13}{52}\cdot\frac{12}{51} = \frac{156}{2652} \approx 0.059$$

5. (a) Throw a large number of similar thumbtacks, or one thumbtack a large number of times, and record the frequency of occurrence of the various outcomes. Assume the thumbtack falls either flat side down (i.e., point up), or tilted (with the point down, resting on the edge of the flat side). (We will assume these are the only two ways the tack can land.)

To estimate the probability the tack lands on its flat side with the point up, find the relative frequency of this occurrence by dividing the number of times this occurred by the total number of thumbtack tosses.

(b) The sample space is the two outcomes flat side down (point up) and tilted (point down).

(c) $P(\text{flat side down, point up}) = \dfrac{340}{500} = 0.68$

$P(\text{tilted, point down}) = 1 - 0.68 = 0.32$

7. (a) possible values for x, the sum of the two dice faces, is 2, 3, 4, 5, 6, 7, 8, 9, 10, 11, and 12

(b) 2:1 and 1
3:1 and 2, or 2 and 1
4:1 and 3, 2 and 2, 3 and 1
5:1 and 4, 2 and 3, 3 and 2, 4 and 1
6:1 and 5, 2 and 4, 3 and 3, 4 and 2, 5 and 1
7:1 and 6, 2 and 5, 3 and 4, 4 and 3, 5 and 2, 6 and 1
8:2 and 6, 3 and 5, 4 and 4, 5 and 3, 6 and 2
9:3 and 6, 4 and 5, 5 and 4, 6 and 3
10:4 and 6, 5 and 5, 6 and 4
11:5 and 6, 6 and 5
12:6 and 6

x	$P(x)$
2	$\frac{1}{36} \approx 0.028$
3	$\frac{2}{36} \approx 0.056$
4	$\frac{3}{36} \approx 0.083$
5	$\frac{4}{36} \approx 0.111$
6	$\frac{5}{36} \approx 0.139$
7	$\frac{6}{36} \approx 0.167$
8	$\frac{5}{36} \approx 0.139$
9	$\frac{4}{36} \approx 0.111$
10	$\frac{3}{36} \approx 0.083$
11	$\frac{2}{36} \approx 0.056$
12	$\frac{1}{36} \approx 0.028$

Where there are (6)(6) = 36 possible, equally likely outcomes (the sums, however, are not equally likely).

9. $C_{8,2} = \frac{8!}{2!6!} = \frac{8 \cdot 7 \cdot 6!}{(2 \cdot 1)6!} = \frac{56}{2} = 28$

11. $3 \cdot 2 \cdot 1 = 6$

13. 5 multiple choice questions, each with 4 possible answers (A, B, C, or D), so 4 answers for first question; and for each of those, 4 answers for the second question; and for each of those, 4 answers for the third question; and for each of those, 4 answers for the fifth question. There are $4 \cdot 4 \cdot 4 \cdot 4 \cdot 4 = 4^5 = 1024$ possible sequences, such as A, D, B, B, C or C, B, A, D, D, etc.

$P(\text{getting the correct sequence}) = \frac{1}{1024} \approx 0.00098$

15. 10 possible numbers per turn of dial; 3 dial turns $10 \cdot 10 \cdot 10 = 1000$ possible combinations

Chapter 5 The Binomial Probability Distribution and Related Topics

Section 5.1

1. **(a)** The number of traffic fatalities can be only a whole number. This is a discrete random variable.

   (b) Distance can assume any value, so this is a continuous random variable.

   (c) Time can take on any value, so this is a continuous random variable.

   (d) The number of ships can be only a whole number. This is a discrete random variable.

   (e) Weight can assume any value, so this is a continuous random variable.

3. **(a)** $\Sigma P(x) = 0.25 + 0.60 + 0.15 = 1.00$

   Yes, this is a valid probability distribution because a probability is assigned to each distinct value of the random variable and the sum of these probabilities is 1.

   (b) $\Sigma P(x) = 0.25 + 0.60 + 0.20 = 1.05$

   No, this is not a probability distribution because the probabilities total to more than 1.

5. **(a)** $\Sigma P(x) = 0.21 + 0.14 + 0.22 + 0.15 + 0.20 + 0.08 = 1.00$

   Yes, this is a valid probability distribution because the events are distinct and the probabilities total to 1.

   (b) Income Distribution ($1000)

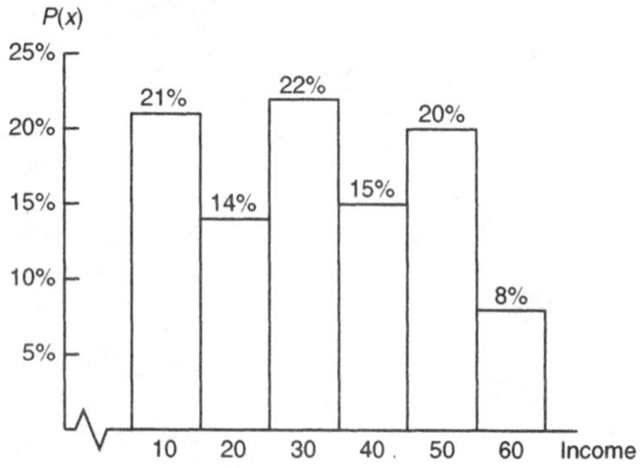

   (c) $\mu = \Sigma x P(x)$

   $= 10(0.21) + 20(0.14) + 30(0.22) + 40(0.15) + 50(0.20) + 60(0.08)$

   $= 32.3$

   (d) $\sigma = \sqrt{\Sigma(x-\mu)^2 P(x)}$

   $= \sqrt{(-22.3)^2(0.21) + (-12.3)^2(0.14) + (-2.3)^2(0.22) + (7.7)^2(0.15) + (17.7)^2(0.20) + (27.7)^2(0.08)}$

   $= \sqrt{259.71}$

   ≈ 16.12

7. (a)

x	f	Relative Frequency	$P(x)$
36	6	6/208	0.029
37	10	10/208	0.048
38	11	11/208	0.053
39	20	20/208	0.096
40	26	26/208	0.125
41	32	32/208	0.154
42	34	34/208	0.163
43	28	28/208	0.135
44	25	25/208	0.120
45	16	16/208	0.077

(b)

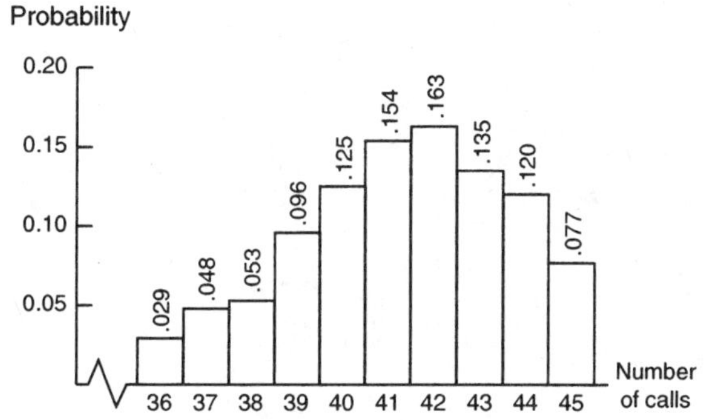

(c) $P(39,40,41,42,43) = P(39)+P(40)+P(41)+P(42)+P(43)$
$= 0.096+0.125+0.154+0.163+0.135$
$= 0.673$

(d) $P(36,37,38,39,40) = P(36)+P(37)+P(38)+P(39)+P(40)$
$= 0.029+0.048+0.053+0.096+0.125$
$= 0.351$

(e) $\mu = \Sigma xP(x)$

$$= 36(0.029)+37(0.048)+38(0.053)+39(0.096)+40(0.125)+41(0.154) +42(0.163)+43(0.135)+44(0.120)+45(0.077)$$

$= 41.288$

(f) $\sigma = \sqrt{\Sigma(x-\mu)^2 P(x)}$

$$= \sqrt{\begin{aligned}&(-5.288)^2(0.029)+(-4.288)^2(0.048)+(-3.288)^2(0.053)+(-2.288)^2(0.096)+(-1.288)^2(0.125)\\&+(-0.288)^2(0.154)+(0.712)^2(0.163)+(1.712)^2(0.135)+(2.712)^2(0.120)+(3.712)^2(0.077)\end{aligned}}$$

$= \sqrt{5.411}$

≈ 2.326

9. (a) Number of Fish Caught in a 6-Hour Period at Pyramid Lake, Nevada

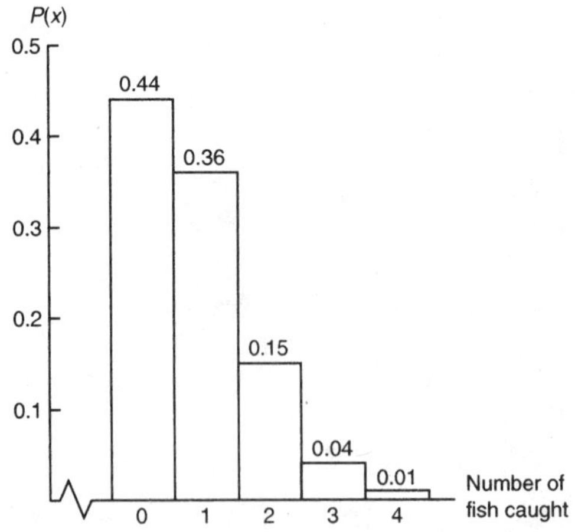

(b) $P(1 \text{ or more}) = 1 - P(0)$
$= 1 - 0.44$
$= 0.56$

(c) $P(2 \text{ or more}) = P(2) + P(3) + P(4 \text{ or more})$
$= 0.15 + 0.04 + 0.01$
$= 0.20$

(d) $\mu = \Sigma xP(x)$
$= 0(0.44) + 1(0.36) + 2(0.15) + 3(0.04) + 4(0.01)$
$= 0.82$

(e) $\sigma = \sqrt{\Sigma(x-\mu)^2 P(x)}$
$= \sqrt{(-0.82)^2 (0.44) + (0.18)^2 (0.36) + (1.18)^2 (0.15) + (2.18)^2 (0.04) + (3.18)^2 (0.01)}$
$= \sqrt{0.8076}$
≈ 0.899

11. (a) $P(\text{win}) = \frac{15}{719} \approx 0.021$

$P(\text{not win}) = \frac{719-15}{719} = \frac{704}{719} \approx 0.979$

(b) Expected earnings $= (\text{value of dinner})(\text{probability of winning})$
$= \$35\left(\frac{15}{719}\right)$
$\approx \$0.73$

Lisa's expected earnings are \$0.73.

contribution $= \$15 - \$0.73 = \$14.27$

Lisa effectively contributed \$14.27 to the hiking club.

13. (a) $P(60 \text{ years}) = 0.01191$

Expected loss = $\$50{,}000(0.01191) = \$595.50$

The expected loss for Big Rock Insurance is \$595.50.

(b)

Probability	Expected Loss
$P(61) = 0.01292$	$\$50{,}000(0.01292) = \$646$
$P(62) = 0.01396$	$\$50{,}000(0.01396) = \$698$
$P(63) = 0.01503$	$\$50{,}000(0.01503) = \$751.50$
$P(64) = 0.01613$	$\$50{,}000(0.01613) = \$806.50$

Expected loss $= \$595.50 + \$646 + \$698 + \$751.50 + \$806.50$
$= \$3497.50$

The total expected loss is \$3497.50.

(c) \$3497.50 + \$700 = \$4197.50

They should charge \$4197.50.

(d) $\$5000 - \$3497.50 = \$1502.50$

They can expect to make \$1502.50.

Comment: losses are usually denoted by negative numbers such as –\$50,000.

15. (a) $W = x_1 - x_2$; $a = 1$, $b = -1$

$$\mu_W = \mu_1 - \mu_2 = 115 - 100 = 15$$

$$\sigma_W^2 = 1^2\sigma_1^2 + (-1)^2\sigma_2^2 = 12^2 + 8^2 = 208$$

$$\sigma_W = \sqrt{\sigma_W^2} = \sqrt{208} \approx 14.4$$

(b) $W = 0.5x_1 + 0.5x_2$; $a = 0.5$, $b = 0.5$

$$\mu_W = 0.5\mu_1 + 0.5\mu_2 = 0.5(115) + 0.5(100) = 107.5$$

$$\sigma_W^2 = (0.5)^2\sigma_1^2 + (0.5)^2\sigma_2^2 = 0.25(12)^2 + 0.25(8)^2 = 52$$

$$\sigma_W = \sqrt{\sigma_W^2} = \sqrt{52} \approx 7.2$$

(c) $L = 0.8x_1 - 2$; $a = -2$, $b = 0.8$

$$\mu_L = -2 + 0.8\mu_1 = -2 + 0.8(115) = 90$$

$$\sigma_L^2 = (0.8)^2\sigma_1^2 = 0.64(12)^2 = 92.16$$

$$\sigma_L = \sqrt{\sigma_L^2} = \sqrt{92.16} = 9.6$$

(d) $L = 0.95x_2 - 5$; $a = -5$, $b = 0.95$

$$\mu_L = -5 + 0.95\mu_2 = -5 + 0.95(100) = 90$$

$$\sigma_L^2 = (0.95)^2\sigma_2^2 = 0.9025(8)^2 = 57.76$$

$$\sigma_L = \sqrt{\sigma_L^2} = \sqrt{57.76} = 7.6$$

17. (a) $W = 0.5x_1 + 0.5x_2$; $a = 0.5$, $b = 0.5$

$\mu_W = 0.5\mu_1 + 0.5\mu_2 = 0.5(50.2) + 0.5(50.2) = 50.2$

$\sigma_W^2 = 0.5^2\sigma_1^2 + 0.5^2\sigma_2^2 = 0.5^2(11.5)^2 + 0.5^2(11.5)^2 = 66.125$

$\sigma_W = \sqrt{\sigma_W^2} = \sqrt{66.125} \approx 8.13$

(b) Single policy (x_1): $\mu_1 = 50.2$

Two policies (W): $\mu_W \approx 50.2$

The means are the same.

(c) Single policy (x_1): $\sigma_1 = 11.5$

Two policies (W): $\sigma_W \approx 8.13$

The standard deviation for two policies is smaller.

(d) Yes, the risk decreases by a factor of $\frac{1}{\sqrt{n}}$ because $\sigma_W = \frac{1}{\sqrt{n}}\sigma$.

Section 5.2

1. A trial is one flip of a fair quarter. Success = head. Failure = tail.

$n = 3$, $p = 0.5$, $q = 1 - 0.5 = 0.5$

(a) $P(3) = C_{3,3}(0.5)^3(0.5)^{3-3}$

$= 1(0.5)^3(0.5)^0$

$= 0.125$

To find this value in Table 3 of Appendix II, use the group in which $n = 3$, the column headed by $p = 0.5$, and the row headed by $r = 3$.

(b) $P(2) = C_{3,2}(0.5)^2(0.5)^{3-2}$

$= 3(0.5)^2(0.5)^1$

$= 0.375$

To find this value in Table 3 of Appendix II, use the group in which $n = 3$, the column headed by $p = 0.5$, and the row headed by $r = 2$.

(c) $P(r \geq 2) = P(2) + P(3)$

$= 0.125 + 0.375$

$= 0.5$

(d) The probability of getting exactly three tails is the same as getting exactly zero heads.

$P(0) = C_{3,0}(0.5)^0(0.5)^{3-0}$

$= 1(0.5)^0(0.5)^3$

$= 0.125$

To find this value in Table 3 of Appendix II, use the group in which $n = 3$, the column headed by $p = 0.5$, and the row headed by $r = 0$.

The results from Table 3 of Appendix II are the same.

In the problems that follow, there are often other ways the solve the problems than those shown. As long as you get the same answer, your method is probably correct.

3. (a) A trial is a man's response to the question, "Would you marry the same woman again?"
Success = a positive response. Failure = a negative response.
$n = 10,\ p = 0.80,\ q = 1 - 0.80 = 0.20$
Using values in Table 3 of Appendix II:

$$\begin{aligned} P(r \geq 7) &= P(7) + P(8) + P(9) + P(10) \\ &= 0.201 + 0.302 + 0.268 + 0.107 \\ &= 0.878 \end{aligned}$$

$$\begin{aligned} P(r \text{ is less than half of } 10) &= P(r < 5) \\ &= P(0) + P(1) + P(2) + P(3) + P(4) \\ &= 0.000 + 0.000 + 0.000 + 0.001 + 0.006 \\ &= 0.007 \end{aligned}$$

(b) A trial is a woman's response to the question, "Would you marry the same man again?"
Success = a positive response. Failure = a negative response.
$n = 10,\ p = 0.5,\ q = 1 - 0.5 = 0.5$
Using values in Table 3 of Appendix II:

$$\begin{aligned} P(r \geq 7) &= P(7) + P(8) + P(9) + P(10) \\ &= 0.117 + 0.044 + 0.010 + 0.001 \\ &= 0.172 \end{aligned}$$

$$\begin{aligned} P(r < 5) &= P(0) + P(1) + P(2) + P(3) + P(4) \\ &= 0.001 + 0.010 + 0.044 + 0.117 + 0.205 \\ &= 0.377 \end{aligned}$$

5. A trial consists of a woman's response regarding her mother-in-law. Success = dislike. Failure = like.
$n = 6,\ p = 0.90,\ q = 1 - 0.90 = 0.10$

(a)
$$\begin{aligned} P(6) &= C_{6,6}(0.90)^6(0.10)^{6-6} \\ &= 1(0.90)^6(0.10)^0 \\ &= 0.531 \end{aligned}$$

(b)
$$\begin{aligned} P(0) &= C_{6,0}(0.90)^0(0.10)^{6-0} \\ &= 1(0.90)^0(0.10)^6 \\ &\approx 0.000 \text{ (to 3 digits)} \end{aligned}$$

(c)
$$\begin{aligned} P(r \geq 4) &= P(4) + P(5) + P(6) \\ &= 0.098 + 0.354 + 0.531 \\ &= 0.983 \end{aligned}$$

(d) $P(r \le 3) = 1 - P(r \ge 4)$
$\approx 1 - 0.983$
$= 0.017$

From the table:
$P(r \le 3) = P(0) + P(1) + P(2) + P(3)$
$= 0.000 + 0.000 + 0.001 + 0.015$
$= 0.016$

7. A trial consists of taking a polygraph examination. Success = pass. Failure = fail.
$n = 9,\ p = 0.85,\ q = 1 - 0.85 = 0.15$

(a) $P(9) = 0.232$

(b) $P(r \ge 5) = P(5) + P(6) + P(7) + P(8) + P(9)$
$= 0.028 + 0.107 + 0.260 + 0.368 + 0.232$
$= 0.995$

(c) $P(r \le 4) = 1 - P(r \ge 5)$
$= 1 - 0.995$
$= 0.005$

From the table:
$P(r \le 4) = P(0) + P(1) + P(2) + P(3) + P(4)$
$= 0.000 + 0.000 + 0.000 + 0.001 + 0.005$
$= 0.006$

The two results should be equal, but because of rounding error, they differ slightly.

(d) All students fail is the same as no students pass.
$P(0) = 0.000$ (to 3 digits)

9. A trial consists of checking the gross receipts of the Green Parrot Italian Restaurant for one business day. Success = gross is over \$2200. Failure = gross is at or below \$2200.
$p = 0.85,\ q = 1 - 0.85 = 0.15$

(a) $n = 7$
$P(r \ge 5) = P(5) + P(6) + P(7)$
$= 0.210 + 0.396 + 0.321$
$= 0.927$

(b) $n = 10$
$P(r \ge 5) = P(5) + P(6) + P(7) + P(8) + P(9) + P(10)$
$= 0.008 + 0.040 + 0.130 + 0.276 + 0.347 + 0.197$
$= 0.998$

(c) $n = 5$
$P(r < 3) = P(0) + P(1) + P(2)$
$= 0.000 + 0.002 + 0.024$
$= 0.026$

(d) $n = 10$

$$P(r < 7) = P(0) + P(1) + P(2) + P(3) + P(4) + P(5) + P(6)$$
$$= 0.000 + 0.000 + 0.000 + 0.000 + 0.001 + 0.008 + 0.040$$
$$= 0.049$$

(e) $n = 7$

$$P(r < 3) = P(0) + P(1) + P(2)$$
$$= 0.000 + 0.000 + 0.001$$
$$= 0.001$$

Yes. If p were really 0.85, then the event of a 7-day period with gross income exceeding \$2200 fewer than 3 days would be very rare. If it happened again, we would suspect that $p = 0.85$ is too high.

11. A trial is catching and releasing a pike. Success = pike dies. Failure = pike lives.
$n = 16$, $p = 0.05$, $q = 1 - 0.05 = 0.95$

(a) $P(0) = 0.440$

(b) $P(r < 3) = P(0) + P(1) + P(2)$
$$= 0.440 + 0.371 + 0.146$$
$$= 0.957$$

(c) All of the fish lived is the same as none of the fish died.
$P(0) = 0.440$

(d) More than 14 fish lived is the same as less than 2 fish died.

$$P(r < 2) = P(0) + P(1)$$
$$= 0.440 + 0.371$$
$$= 0.811$$

13. (a) A trial consists of using the Myers-Briggs instrument to determine if a person in marketing is an extrovert. Success = extrovert. Failure = not extrovert.
$n = 15$, $p = 0.75$, $q = 1 - 0.75 = 0.25$

$$P(r \geq 10) = P(10) + P(11) + P(12) + P(13) + P(14) + P(15)$$
$$= 0.165 + 0.225 + 0.225 + 0.156 + 0.067 + 0.013$$
$$= 0.851$$

$$P(r \geq 5) = P(5) + P(6) + P(7) + P(8) + P(9) + P(r \geq 10)$$
$$= 0.001 + 0.003 + 0.013 + 0.039 + 0.092 + 0.851$$
$$= 0.999$$

$$P(15) = 0.013$$

(b) A trial consists of using the Myers-Briggs instrument to determine if a computer programmer is an introvert. Success = introvert. Failure = not introvert.

$n = 5,\ p = 0.60,\ q = 1 - 0.60 = 0.40$

$$P(0) = 0.010$$

$$\begin{aligned} P(r \ge 3) &= P(3) + P(4) + P(5) \\ &= 0.346 + 0.259 + 0.078 \\ &= 0.683 \end{aligned}$$

$$P(5) = 0.078$$

15. A trial is checking the development of hypertension in patients with diabetes. Success = yes. Failure = no.

$n = 10,\ p = 0.40,\ q = 1 - 0.40 = 0.60$

(a) $P(0) = 0.006$

(b) $$\begin{aligned} P(r < 5) &= P(0) + P(1) + P(2) + P(3) + P(4) \\ &= 0.006 + 0.040 + 0.121 + 0.215 + 0.251 \\ &= 0.633 \end{aligned}$$

A trial is checking the development of an eye disease in patients with diabetes. Success = yes. Failure = no.

$n = 10,\ p = 0.30,\ q = 1 - 0.30 = 0.70$

(c) $$\begin{aligned} P(r \le 2) &= P(0) + P(1) + P(2) \\ &= 0.028 + 0.121 + 0.233 \\ &= 0.382 \end{aligned}$$

(d) At least 6 will never develop a related eye disease is the same as at most 4 will never develop a related eye disease.

$$\begin{aligned} P(r \le 4) &= P(0) + P(1) + P(2) + P(3) + P(4) \\ &= 0.028 + 0.121 + 0.233 + 0.267 + 0.200 \\ &= 0.849 \end{aligned}$$

17. A trial consists of the response of adults regarding their concern that Social Security numbers are used for general identification. Success = concerned that SS numbers are being used for identification. Failure = not concerned that SS numbers are being used for identification.

$n = 8,\ p = 0.53,\ q = 1 - 0.53 = 0.47$

(a) $$\begin{aligned} P(r \le 5) &= P(0) + P(1) + P(2) + P(3) + P(4) + P(5) \\ &= 0.002381 + 0.021481 + 0.084781 + 0.191208 + 0.269521 + 0.243143 \\ &= 0.812515 \end{aligned}$$

$P(r \le 5) = 0.81251$ from the cumulative probability is the same, truncated to 5 digits.

(b) $P(r>5)=P(6)+P(7)+P(8)$

$$=0.137091+0.044169+0.006726$$
$$=0.187486$$

$$P(r>5)=1-P(r\leq 5)$$
$$=1-0.81251$$
$$=0.18749$$

Yes, this is the same result rounded to 5 digits.

19. A trial consists of determining the kind of stone in a chipped stone tool.

(a) $n = 11$; Success = obsidian. Failure = not obsidian.

$p=0.15,\ q=1-0.15=0.85$

$$P(r\geq 3)=1-P(r\leq 2)$$
$$=1-\left[P(0)+P(1)+P(2)\right]$$
$$=1-(0.167+0.325+0.287)$$
$$=1-0.779$$
$$=0.221$$

(b) $n = 5$; Success = basalt. Failure = not basalt.

$p=0.55,\ q=1-0.55=0.45$

$$P(r\geq 2)=P(2)+P(3)+P(4)+P(5)$$
$$=0.276+0.337+0.206+0.050$$
$$=0.869$$

(c) $n = 10$; Success = neither obsidian nor basalt. Failure = either obsidian or basalt. The two outcomes, tool is obsidian or tool is basalt, are mutually exclusive. Therefore, *P*(obsidian *or* basalt) = 0.55 + 0.15 = 0.70. *P*(neither obsidian nor basalt) = 1 – 0.70 = 0.30. Therefore, $p = 0.30$, $q = 1 - 0.30 = 0.70$.

$$P(r\geq 4)=P(4)+P(5)+P(6)+P(7)+P(8)+P(9)+P(10)$$
$$=0.200+0.103+0.037+0.009+0.001+0.000+0.000$$
$$=0.350$$

Note that the phrase "neither obsidian nor basalt" is the English translation of the math phrase "not (obsidian or basalt)" and therefore, it describes the complement of the event "obsidian or basalt." Using $P(\overline{A})=1-P(A)$, we get p = *P*(Success) = 0.30.

21. **(a)** $p=0.30,\ P(3)=0.132$

$p=0.70,\ P(2)=0.132$

They are the same.

(b) $p=0.30,\ P(r\geq 3)=0.132+0.028+0.002=0.162$

$p=0.70,\ P(r\leq 2)=0.002+0.028+0.132=0.162$

They are the same.

(c) $p=0.30,\ P(4)=0.028$

$p=0.70,\ P(1)=0.028$

$r=1$

(d) The column headed by $p = 0.80$ is symmetrical with the one headed by $p = 0.20$.

23. (a) $n = 8; p = 0.65$

$$P(6 \le r, \text{ given } 4 \le r) = \frac{P(6 \le r \text{ and } 4 \le r)}{P(4 \le r)}$$
$$= \frac{P(6 \le r)}{P(4 \le r)}$$
$$= \frac{P(6) + P(7) + P(8)}{P(4) + P(5) + P(6) + P(7) + P(8)}$$
$$= \frac{0.259 + 0.137 + 0.032}{0.188 + 0.279 + 0.259 + 0.137 + 0.032}$$
$$= \frac{0.428}{0.895}$$
$$= 0.478$$

(b) $n = 10; p = 0.65$

$$P(8 \le r, \text{ given } 6 \le r) = \frac{P(8 \le r \text{ and } 6 \le r)}{P(6 \le r)}$$
$$= \frac{P(8 \le r)}{P(6 \le r)}$$
$$= \frac{P(8) + P(9) + P(10)}{P(6) + P(7) + P(8) + P(9) + P(10)}$$
$$= \frac{0.176 + 0.072 + 0.014}{0.238 + 0.252 + 0.176 + 0.072 + 0.014}$$
$$= \frac{0.262}{0.752}$$
$$= 0.348$$

(c) Essay.

(d) Let event $A = 6 \le r$ and event $B = 4 \le r$ in the formula.

Section 5.3

1. (a) Binomial Distribution
The distribution is symmetrical.

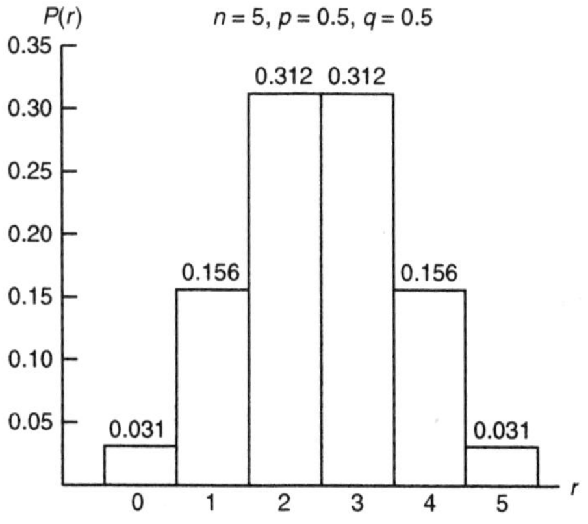

(b) Binomial Distribution
The distribution is skewed right.

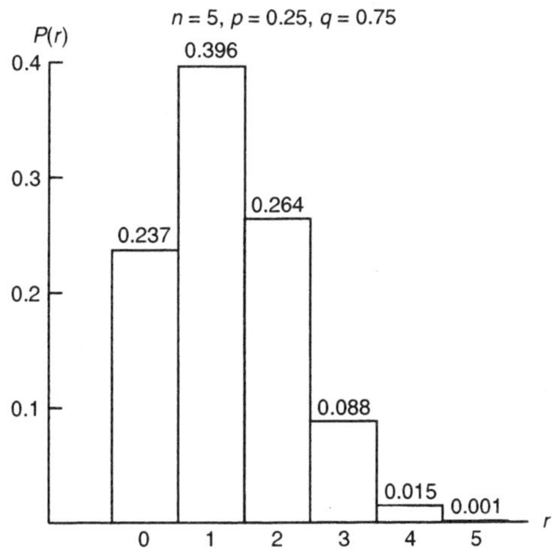

(c) Binomial Distribution
The distribution is skewed left.

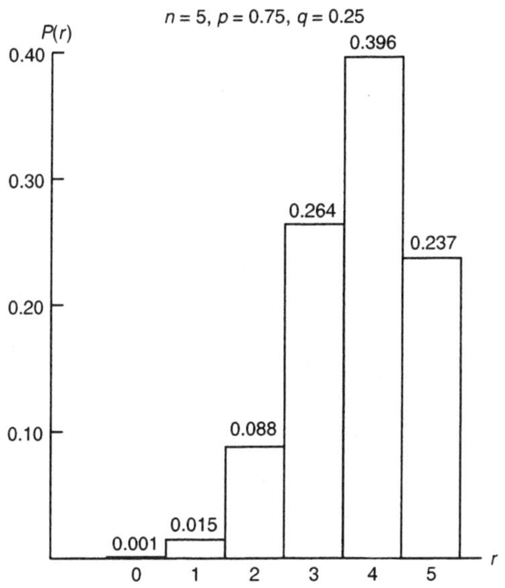

(d) The distributions are mirror images of one another.

(e) The distribution would be skewed left for $p = 0.73$ because the more likely number of successes are to the right of the middle.

3. The probabilities can be taken directly from Table 3 in Appendix II.

(a) $n = 10,\ p = 0.80$

Households with Children Under 2 That Buy Film

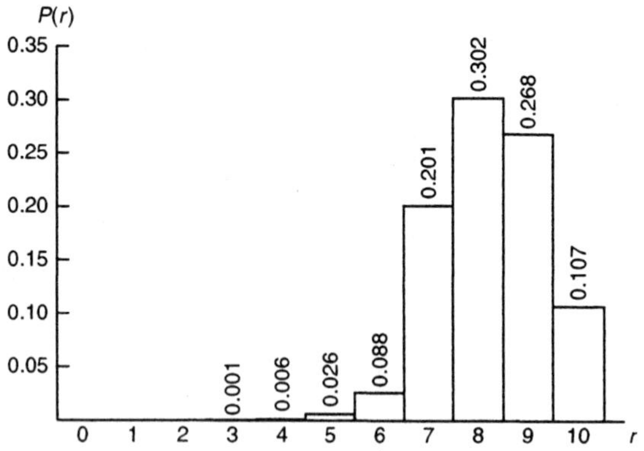

$\mu = np = 10(0.8) = 8$

$\sigma = \sqrt{npq} = \sqrt{10(0.8)(0.2)} \approx 1.26$

(b) $n = 10,\ p = 0.5$

Households with No Children Under 21 That Buy Film

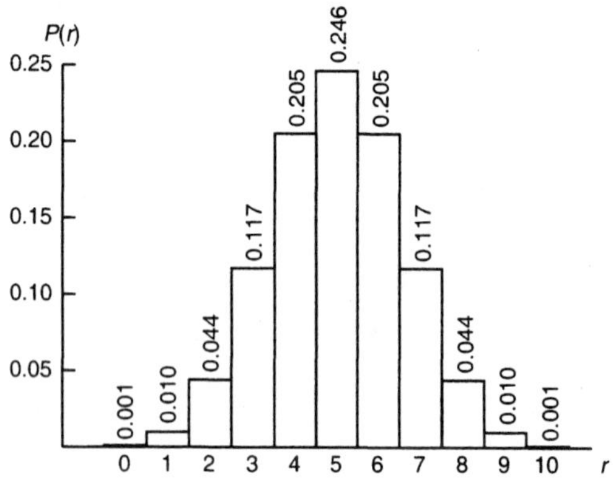

$\mu = np = 10(0.5) = 5$

$\sigma = \sqrt{npq} = \sqrt{10(0.5)(0.5)} \approx 1.58$

(c) Yes; since the graph in part (a) is skewed left, it supports the claim that more households buy film that have children under 2 years than households that have no children under 21 years.

5. (a) $n = 6,\ p = 0.70$

The probabilities can be taken directly from Table 3 in Appendix II.

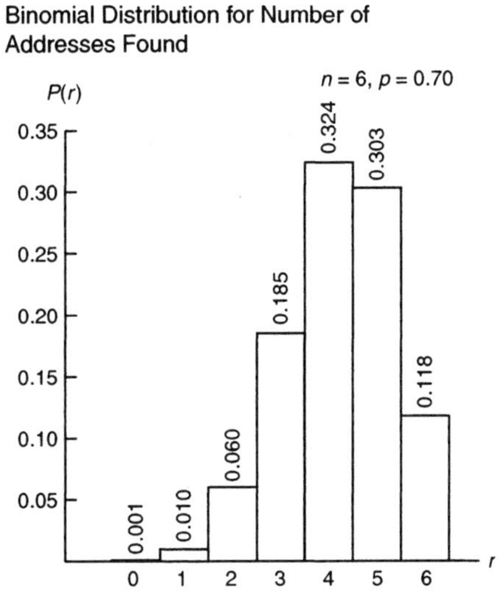

(b) $\mu = np = 6(0.70) = 4.2$

$$\sigma = \sqrt{npq} = \sqrt{6(0.70)(0.30)} \approx 1.122$$

The expected number of friends for whom addresses will be found is 4.2.

(c) Find n such that $P(r \geq 2) = 0.97$.

Try $n = 5$.

$$\begin{aligned} P(r \geq 2) &= P(2) + P(3) + P(4) + P(5) \\ &= 0.132 + 0.309 + 0.360 + 0.168 \\ &= 0.969 \\ &\approx 0.97 \end{aligned}$$

You would have to submit 5 names to be 97% sure that at least two addresses will be found.

If you solve this problem as

$$\begin{aligned} P(r \geq 2) &= 1 - P(r < 2) \\ &= 1 - [P(r = 0) + P(r = 1)] \\ &= 1 - 0.002 - 0.028 = 0.97, \end{aligned}$$

the answers differ due to rounding error in the table.

7. (a) $n = 7,\ p = 0.20$

The probabilities can be taken directly from Table 3 in Appendix II.

Binomial Distribution for Number of Illiterate People

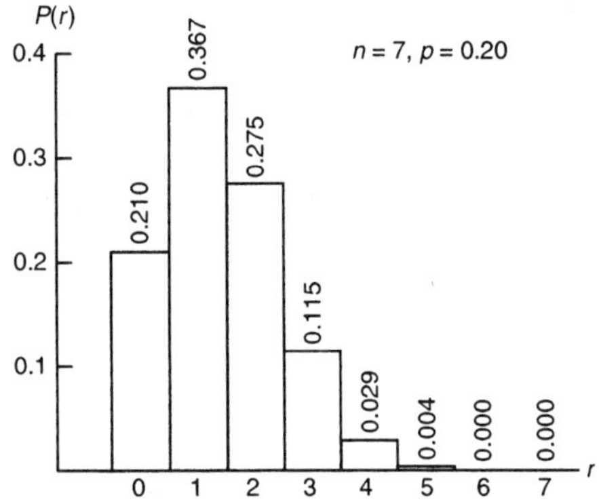

(b) $\mu = np = 7(0.20) = 1.4$

$$\sigma = \sqrt{npq} = \sqrt{7(0.20)(0.80)} \approx 1.058$$

The expected number of people in this sample who are illiterate is 1.4.

(c) Let success = literate and $p = 0.80$.
Find n such that

$$P(r \geq 7) = 0.98.$$

Try $n = 12$.

$$P(r \geq 7) = P(7) + P(8) + P(9) + P(10) + P(11) + P(12)$$
$$= 0.053 + 0.133 + 0.236 + 0.283 + 0.206 + 0.069$$
$$= 0.98$$

You would need to interview 12 people to be 98% sure that at least seven of these people are not illiterate.

9. (a) $n = 8,\ p = 0.25$

The probabilities can be taken directly from Table 3 in Appendix II.

Binomial Distribution for Number of
Gullible Customers

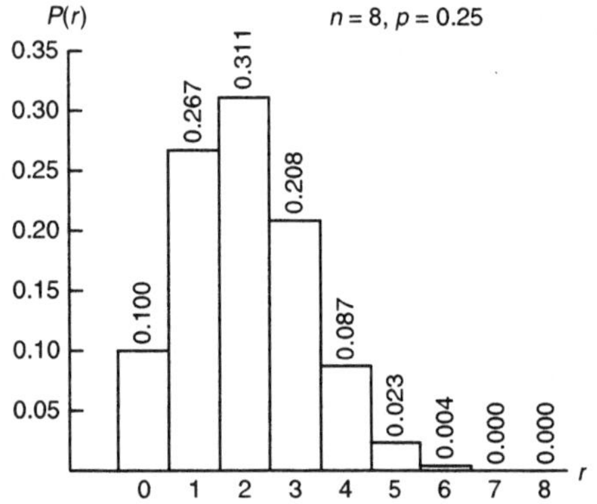

(b) $\mu = np = 8(0.25) = 2$

$$\sigma = \sqrt{npq} = \sqrt{8(0.25)(0.75)} \approx 1.225$$

The expected number of people in this sample who believe the product is improved is 1.4.

(c) Find n such that

$$P(r \geq 1) = 0.99.$$

Try $n = 16$.

$$\begin{aligned} P(r \geq 1) &= 1 - P(0) \\ &= 1 - 0.01 \\ &= 0.99 \end{aligned}$$

Sixteen people are needed in the marketing study to be 99% sure that at least one person believes the product to be improved.

11. $p = 0.40$

Find n such that

$$P(r \geq 5) = 0.95.$$

Try $n = 20$.

$$\begin{aligned} P(r \geq 5) &= 1 - [P(0) + P(1) + P(2) + P(3) + P(4)] \\ &= 1 - (0.000 + 0.000 + 0.003 + 0.012 + 0.035) \\ &= 1 - 0.05 \\ &= 0.95 \end{aligned}$$

He must make 20 sales calls to be 95% sure of meeting the quota.

13. $p = 0.10$

Find n such that

$$P(r \geq 1) = 0.90$$

From a calculator or a computer, we determine $n = 22$ gives $P(r \geq 1) = 0.9015$.

15. **(a)** Since success = not a repeat offender, then $p = 0.75$.

From Table 3 in Appendix II and $n = 4$, $p = 0.75$.

r	0	1	2	3	4
$P(r)$	0.004	0.047	0.211	0.422	0.316

(b) Binomial Distribution for Number of Parolees Who Do Not Become Repeat Offenders

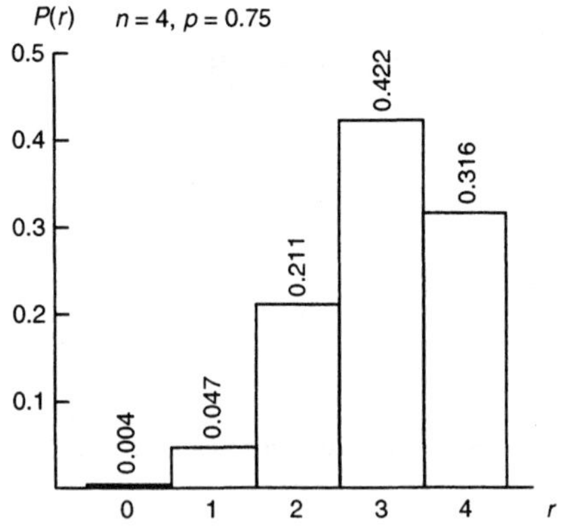

(c) $\mu = np = 4(0.75) = 3$

The expected number of parolees in Alice's group who will not be repeat offenders is 3.

$$\sigma = \sqrt{npq} = \sqrt{4(0.75)(0.25)} \approx 0.866$$

(d) Find n such that

$$P(r \geq 3) = 0.98.$$

Try $n = 7$.

$$P(r \geq 3) = P(3) + P(4) + P(5) + P(6) + P(7)$$
$$= 0.058 + 0.173 + 0.311 + 0.311 + 0.133$$
$$= 0.986$$

This is slightly higher than needed, but $n = 6$ yields $P(r \geq 3) = 0.963$.

Alice should have a group of 7 to be about 98% sure three or more will not become repeat offenders.

17. (a) Let success = available, then $p = 0.75$, $n = 12$.

$$P(12) = 0.032$$

(b) Let success = not available, then $p = 0.25$, $n = 12$.

$$P(r \geq 6) = P(6) + P(7) + P(8) + P(9) + P(10) + P(11) + P(12)$$
$$= 0.040 + 0.011 + 0.002 + 0.000 + 0.000 + 0.000 + 0.000$$
$$= 0.053$$

(c) $n = 12, p = 0.75$

$$\mu = np = 12(0.75) = 9$$

The expected number of those available to serve on the jury is 9.

$$\sigma = \sqrt{npq} = \sqrt{12(0.75)(0.25)} = 1.5$$

(d) $p = 0.75$

Find n such that

$$P(r \geq 12) = 0.959.$$

Try $n = 20$.

$$\begin{aligned} P(r \geq 12) &= P(12) + P(13) + P(14) + P(15) + P(16) + P(17) + P(18) + P(19) + P(20) \\ &= 0.061 + 0.112 + 0.169 + 0.202 + 0.190 + 0.134 + 0.067 + 0.021 + 0.003 \\ &= 0.959 \end{aligned}$$

The jury commissioner must contact 20 people to be 95.9% sure of finding at least 12 people who available to serve.

19. Let success = case solved, then $p = 0.2$, $n = 6$.

(a) $P(0) = 0.262$

(b)
$$\begin{aligned} P(r \geq 1) &= 1 - P(0) \\ &= 1 - 0.262 \\ &= 0.738 \end{aligned}$$

(c) $\mu = np = 6(0.20) = 1.2$

The expected number of crimes that will be solved is 1.2.

$$\sigma = \sqrt{npq} = \sqrt{6(0.20)(0.80)} \approx 0.98$$

(d) Find n such that

$$P(r \geq 1) = 0.90.$$

Try $n = 11$.

$$\begin{aligned} P(r \geq 1) &= 1 - P(0) \\ &= 1 - 0.086 \\ &= 0.914 \end{aligned}$$

[Note: For $n = 10$, $P(r \geq 1) = 0.893$.]

The police must investigate 11 property crimes before they can be at least 90% sure of solving one or more cases.

21. (a) Japan: $n = 7, p = 0.95$

$$P(7) = 0.698$$

United States: $n = 7, p = 0.60$

$$P(7) = 0.028$$

(b) Japan: $n = 7, p = 0.95$

$$\mu = np = 7(0.95) = 6.65$$

$$\sigma = \sqrt{npq} = \sqrt{7(0.95)(0.05)} \approx 0.58$$

United States: $n = 7, p = 0.60$

$$\mu = np = 7(0.60) = 4.2$$

$$\sigma = \sqrt{npq} = \sqrt{7(0.60)(0.40)} \approx 1.30$$

The expected number of verdicts in Japan is 6.65 and in the United States is 4.2.

(c) United States: $p = 0.60$

Find n such that

$$P(r \ge 2) = 0.99.$$

Try $n = 8$.

$$\begin{aligned} P(r \ge 2) &= 1 - [P(0) + P(1)] \\ &= 1 - (0.001 + 0.008) \\ &= 0.991 \end{aligned}$$

Japan: $p = 0.95$

Find n such that

$$P(r \ge 2) = 0.99.$$

Try $n = 3$.

$$\begin{aligned} P(r \ge 2) &= P(2) + P(3) \\ &= 0.135 + 0.857 \\ &= 0.992 \end{aligned}$$

Cover 8 trials in the U.S. and 3 trials in Japan.

23. (a) $p = 0.40$

Find n such that

$$P(r \ge 1) = 0.99.$$

Try $n = 9$.

$$\begin{aligned} P(r \ge 1) &= 1 - P(0) \\ &= 1 - 0.010 \\ &= 0.990 \end{aligned}$$

The owner must answer 9 inquiries to be 99% sure of renting at least one room.

(b) $n = 25, p = 0.40$

$\mu = np = 25(0.40) = 10$

The expected number is 10 room rentals.

Section 5.4

1. (a) Geometric probability distribution, $p = 0.77$.

$$P(n) = p(1-p)^{n-1}$$

$$P(n) = (0.77)(0.23)^{n-1}$$

(b) $P(1) = (0.77)(0.23)^{1-1}$

$$\begin{aligned} &= (0.77)(0.23)^0 \\ &= 0.77 \end{aligned}$$

(c) $P(2) = (0.77)(0.23)^{2-1}$

$$\begin{aligned} &= (0.77)(0.23)^1 \\ &= 0.1771 \end{aligned}$$

(d) $P(3 \text{ or more tries}) = 1 - P(1) - P(2)$
$= 1 - 0.77 - 0.1771$
$= 0.0529$

(e) $\mu = \dfrac{1}{p} = \dfrac{1}{0.77} \approx 1.29$

The expected number is 1 attempt.

3. (a) Geometric probability distribution, $p = 0.05$.

$$P(n) = p(1-p)^{n-1}$$
$$P(n) = (0.05)(0.95)^{n-1}$$

(b) $P(5) = (0.05)(0.95)^{5-1}$
$= (0.05)(0.95)^4$
≈ 0.0407

(c) $P(10) = (0.05)(0.95)^{10-1}$
$= (0.05)(0.95)^9$
≈ 0.0315

(d) $P(\text{more than } 3) = 1 - P(1) - P(2) - P(3)$
$= 1 - 0.05 - (0.05)(0.95) - (0.05)(0.95)^2$
$= 1 - 0.05 - 0.0475 - 0.0451$
$= 0.8574$

(e) $\mu = \dfrac{1}{p} = \dfrac{1}{0.05} \approx 20$

The expected number is 20 pot shards.

5. (a) Geometric probability distribution, $p = 0.71$.

$$P(n) = p(1-p)^{n-1}$$
$$P(n) = (0.71)(0.29)^{n-1}$$

(b) $P(1) = (0.71)(0.29)^{1-1} = 0.71$

$P(2) = (0.71)(0.29)^{2-1} = 0.2059$

$P(n \geq 3) = 1 - P(1) - P(2)$
$= 1 - 0.71 - 0.2059$
$= 0.0841$

(c) $P(n)=(0.83)(0.17)^{n-1}$

$$P(1)=(0.83)(0.17)^{1-1}=0.83$$

$$P(2)=(0.83)(0.17)^{2-1}=0.1411$$

$$\begin{aligned}P(n\geq 3)&=1-P(1)-P(2)\\&=1-0.83-0.1411\\&=0.0289\end{aligned}$$

7. (a) Geometric probability distribution, $p = 0.30$.

$$P(n)=p(1-p)^{n-1}$$

$$P(n)=(0.30)(0.70)^{n-1}$$

(b) $P(3)=(0.30)(0.70)^{3-1}=0.147$

(c)
$$\begin{aligned}P(n>3)&=1-P(1)-P(2)-P(3)\\&=1-0.30-(0.30)(0.70)-0.147\\&=1-0.30-0.21-0.147\\&=0.343\end{aligned}$$

(d) $\mu=\frac{1}{p}=\frac{1}{0.30}=3.33$

The expected number is 3 trips.

9. (a) The Poisson distribution would be a good choice because frequency of initiating social grooming is a relatively rare occurrence. It is reasonable to assume that the events are independent and the variable is the number of times that one otter initiates social grooming in a fixed time interval.

$$\lambda=\frac{1.7}{10\text{ min}}\cdot\frac{3}{3}=\frac{5.1}{30\text{ min}};\ \lambda=5.1\text{ per }30\text{ min interval}$$

$$P(r)=\frac{e^{-\lambda}\lambda^r}{r!}$$

$$P(r)=\frac{e^{-5.1}(5.1)^r}{r!}$$

(b) $P(4)=\frac{e^{-5.1}(5.1)^4}{4!}\approx 0.1719$

$$P(5)=\frac{e^{-5.1}(5.1)^5}{5!}\approx 0.1753$$

$$P(6)=\frac{e^{-5.1}(5.1)^6}{6!}\approx 0.1490$$

(c)
$$\begin{aligned}P(r\geq 4)&=1-P(0)-P(1)-P(2)-P(3)\\&=1-0.0061-0.0311-0.0793-0.1348\\&=0.7487\end{aligned}$$

(d) $P(r<4)=P(0)+P(1)+P(2)+P(3)$

$=0.0061-0.0311-0.0793+0.1348$

$=0.2513$

or

$P(r<4)=1-P(r\geq 4)$

$=1-0.7487$

$=0.2513$

11. (a) Essay. Answer could include:

The Poisson distribution would be a good choice because frequency of births is a relatively rare occurrence. It is reasonable to assume that the events are independent and the variable is the number of births (or deaths) for a community of a given population size.

(b) For 1000 people, $\lambda=16$ births; $\lambda=8$ deaths

By Table 4 in Appendix II:

$P(10\text{ births})=0.0341$

$P(10\text{ deaths})=0.0993$

$P(16\text{ births})=0.0992$

$P(16\text{ deaths})=0.0045$

(c) For 1500 people,

$$\lambda=\frac{16}{1000}\cdot\frac{1.5}{1.5}=\frac{24}{1500};\ \lambda=24\text{ births per 1500 people}$$

$$\lambda=\frac{8}{1000}\cdot\frac{1.5}{1.5}=\frac{12}{1500};\ \lambda=12\text{ deaths per 1500 people}$$

By Table 4 or a calculator:

$P(10\text{ births})=0.00066$

$P(10\text{ deaths})=0.1048$

$P(16\text{ births})=0.02186$

$P(16\text{ deaths})=0.0543$

(d) For 750 people,

$$\lambda=\frac{16}{1000}\cdot\frac{0.75}{0.75}=\frac{12}{750};\ \lambda=12\text{ births per 750 people}$$

$$\lambda=\frac{8}{1000}\cdot\frac{0.75}{0.75}=\frac{6}{750};\ \lambda=6\text{ deaths per 750 people}$$

$P(10\text{ births})=0.1048$

$P(10\text{ deaths})=0.0413$

$P(16\text{ births})=0.0543$

$P(16\text{ deaths})=0.0003$

13. (a) Essay. Answer could include:

The Poisson distribution would be a good choice because frequency of gale-force winds is a relatively rare occurrence. It is reasonable to assume that the events are independent and the variable is the number of gale-force winds in a given time interval.

(b) $\lambda = \frac{1}{60 \text{ hr}} \cdot \frac{1.8}{1.8} = \frac{1.8}{108 \text{ hours}};$

$\lambda = 1.8$ per 108 hours

From Table 4 in Appendix II:

$$P(2) = 0.2678$$
$$P(3) = 0.1607$$
$$P(4) = 0.0723$$
$$\begin{aligned} P(r<2) &= P(0) + P(1) \\ &= 0.1653 + 0.2975 \\ &= 0.4628 \end{aligned}$$

(c) $\lambda = \frac{1}{60 \text{ hr}} \cdot \frac{3}{3} = \frac{3}{180 \text{ hours}};$

$\lambda = 3$ per 180 hours

$$P(3) = 0.2240$$
$$P(4) = 0.1680$$
$$P(5) = 0.1008$$
$$\begin{aligned} P(r<2) &= P(0) + P(1) + P(2) \\ &= 0.0498 + 0.1494 + 0.2240 \\ &= 0.4232 \end{aligned}$$

15. (a) Essay. Answer could include:

The Poisson distribution would be a good choice because frequency of commercial building sales is a relatively rare occurrence. It is reasonable to assume that the events are independent and the variable is the number of buildings sold in a given time interval.

(b) $\lambda = \frac{8}{275 \text{ days}} \cdot \frac{\frac{12}{55}}{\frac{12}{55}} \approx \frac{\frac{96}{55}}{60 \text{ days}};$

$\lambda = \frac{96}{55} \approx 1.7$ per 60 days

From Table 4 in Appendix II:

$$P(0) = 0.1827$$
$$P(1) = 0.3106$$
$$\begin{aligned} P(r \geq 2) &= 1 - P(0) - P(1) \\ &= 1 - 0.1827 - 0.3106 \\ &= 0.5067 \end{aligned}$$

(c) $\lambda = \dfrac{8}{275 \text{ days}} \cdot \dfrac{\frac{18}{55}}{\frac{18}{55}} \approx \dfrac{2.6}{90 \text{ days}}$;

$\lambda \approx 2.6$ per 90 days

$$P(0) = 0.0743$$
$$P(2) = 0.2510$$
$$\begin{aligned} P(r \geq 3) &= 1 - P(0) - P(1) - P(2) \\ &= 1 - 0.0743 - 0.1931 - 0.2510 \\ &= 0.4816 \end{aligned}$$

17. (a) Essay. Answer could include:

The problem satisfies the conditions for a binomial experiment with

n large, $n = 1000$, and p small, $p = \frac{1}{569} \approx 0.0018$.
$np \approx 1000(0.0018) = 1.8 < 10$.

The Poisson distribution would be a good approximation to the binomial.

$$\lambda = np \approx 1.8$$

(b) From Table 4 in Appendix II,

$$P(0) = 0.1653$$

(c) $$\begin{aligned} P(r > 1) &= 1 - P(0) - P(1) \\ &= 1 - 0.1653 - 0.2975 \\ &= 0.5372 \end{aligned}$$

(d) $$\begin{aligned} P(r > 2) &= P(r > 1) - P(2) \\ &= 0.5372 - 0.2678 \\ &= 0.2694 \end{aligned}$$

(e) $$\begin{aligned} P(r > 3) &= P(r > 2) - P(3) \\ &= 0.2694 - 0.1607 \\ &= 0.1087 \end{aligned}$$

19. (a) Essay. Answer could include:

The problem satisfies the conditions for a binomial experiment with n large, $n = 175$, and p small, $p = 0.005$. $np = (175)(0.005) = 0.875 < 10$. The Poisson distribution would be a good approximation to the binomial. $n = 175$, $p = 0.005$, $\lambda = np = 0.9$.

(b) From Table 4 in Appendix II,

$$P(0) = 0.4066$$

(c) $$\begin{aligned} P(r \geq 1) &= 1 - P(0) \\ &= 1 - 0.4066 \\ &= 0.5934 \end{aligned}$$

(d) $P(r \geq 2) = P(r \geq 1) - P(1)$
$= 0.5934 - 0.3659$
$= 0.2275$

21. (a) $n = 100, p = 0.02, r = 2$

$$P(r) = C_{n,r}p^r(1-p)^{n-r}$$
$$P(2) = C_{100,2}(0.02)^2(0.98)^{100-2}$$
$$= 4950(0.0004)(0.1381)$$
$$= 0.2734$$

(b) $\lambda = np = 100(0.02) = 2$

From Table 4 in Appendix II,

$$P(2) = 0.2707$$

(c) The approximation is correct to two decimal places.

(d) $n = 100; p = 0.02; r = 3$

By the formula for the binomial distribution,

$$P(3) = C_{100,3}(0.02)^3(0.98)^{100-3}$$
$$= 161{,}700(0.000008)(0.1409)$$
$$= 0.1823$$

By the Poisson approximation, $\lambda = 3$, $P(3) = 0.1804$. The approximation is correct to two decimal places.

23. (a) The Poisson distribution would be a good choice because hail storms in Western Kansas is a relatively rare occurrence. It is reasonable to assume that the events are independent and the variable is the number of hail storms in a fixed square mile area.

$$\lambda = \frac{2.1}{5} \cdot \frac{\frac{8}{5}}{\frac{8}{5}} = \frac{2.1\left(\frac{8}{5}\right)}{8} \approx \frac{3.4}{8}$$

$\lambda = 3.4$ storms per 8 square miles

(b) $$P(r \geq 4, \textit{given } r \geq 2) = \frac{P(r \geq 4 \text{ and } r \geq 2)}{P(r \geq 2)}$$
$$= \frac{P(r \geq 4)}{P(r \geq 2)}$$
$$= \frac{1 - P(r \leq 3)}{1 - P(r \leq 1)}$$
$$= \frac{1 - (0.0334 + 0.1135 + 0.1929 + 0.2186)}{1 - (0.0334 + 0.1135)}$$
$$= \frac{0.4416}{0.8531}$$
$$= 0.5176$$

(c) $$P(r < 6, \text{ given } r \geq 3) = \frac{P(r < 6 \text{ and } r \geq 3)}{P(r \geq 3)}$$
$$= \frac{P(r=3)+P(r=4)+P(r=5)}{1-P(r \leq 2)}$$
$$= \frac{0.2186+0.1858+0.1264}{1-(0.0334+0.1135+0.1929)}$$
$$= \frac{0.5308}{0.6602}$$
$$= 0.8040$$

25. (a) We have binomial trials for which the probability of success is $p = 0.80$ and failure is $q = 0.20$; $k = 12$ is a fixed whole number ≥ 1; n is a random variable representing the number of contacts needed to get the 12th sale.

$$P(n) = C_{n-1,\, k-1} p^k q^{n-k}$$
$$P(n) = C_{n-1,\, 11}(0.80^{12})(0.20^{n-12})$$

(b) $$P(12) = C_{11,\, 11}(0.80^{12})(0.20^{0}) \approx 0.0687$$
$$P(13) = C_{12,\, 11}(0.80^{12})(0.20^{1}) \approx 0.1649$$
$$P(14) = C_{13,\, 11}(0.80^{12})(0.20^{2}) \approx 0.2144$$

(c) $$P(12 \leq n \leq 14) = P(12)+P(13)+P(14)$$
$$= 0.0687+0.1649+0.2144$$
$$= 0.4480$$

(d) $$P(n > 14) = 1-P(n \leq 14)$$
$$= 1-P(12 \leq n \leq 14)$$
$$= 1-0.4480$$
$$= 0.5520$$

(e) $$\mu = \frac{k}{p} = \frac{12}{0.80} = 15$$
$$\sigma = \frac{\sqrt{kq}}{p} = \frac{\sqrt{12(0.20)}}{0.80} \approx 1.94$$

The expected contact in which the 12th sale will occur is 15 with a standard deviation of 1.94.

27. Essay.

Chapter 5 Review

1. (a) $\mu = \Sigma xP(x)$

$= 18.5(0.127) + 30.5(0.371) + 42.5(0.285) + 54.5(0.215) + 66.5(0.002)$

$= 37.628$

≈ 37.63

The expected lease term is about 38 months.

$\sigma = \sqrt{\Sigma(x-\mu)^2 P(x)}$

$= \sqrt{(-19.13)^2(0.127) + (-7.13)^2(0.371) + (4.87)^2(0.285) + (16.87)^2(0.215) + (28.87)^2(0.002)}$

$\approx \sqrt{134.95}$

≈ 11.6 (using $\mu = 37.63$ in the calculations)

(b) Leases in Months

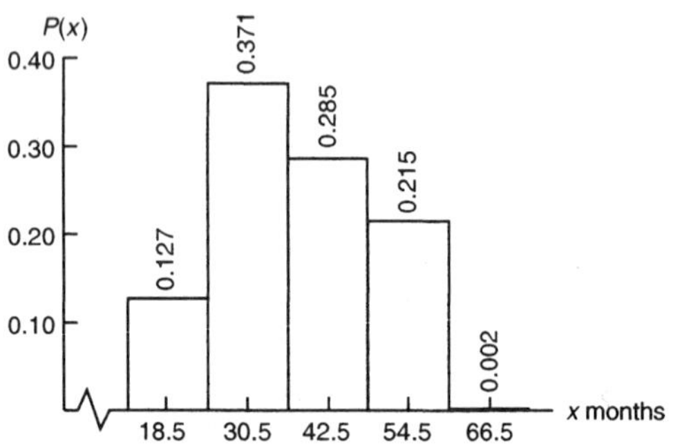

3. This is a binomial experiment with 10 trials. A trial consists of a claim.

Success = submitted by a male under 25 years of age.
Failure = not submitted by a male under 25 years of age.

(a) The probabilities can be taken directly from Table 3 in Appendix II, $n = 10, p = 0.55$.

Claimants Under 25

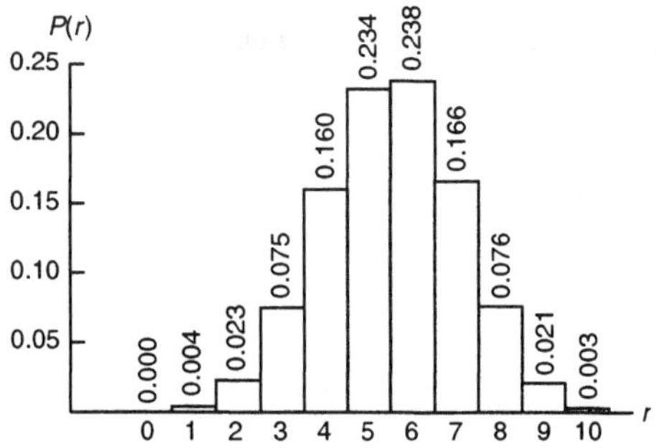

(b) $P(r \geq 6) = P(6) + P(7) + P(8) + P(9) + P(10)$

$= 0.238 + 0.166 + 0.076 + 0.021 + 0.003$

$= 0.504$

(c) $\mu = np = 10(0.55) = 5.5$

The expected number of claims made by males under age 25 is 5.5.

$\sigma = \sqrt{npq} = \sqrt{10(0.55)(0.45)} \approx 1.57$

5. $n = 16, p = 0.50$

(a) $P(r \geq 12) = P(12) + P(13) + P(14) + P(15) + P(16)$

$= 0.028 + 0.009 + 0.002 + 0.000 + 0.000$

$= 0.039$

(b) $P(r \leq 7) = P(0) + P(1) + P(2) + P(3) + P(4) + P(5) + P(6) + P(7)$

$= 0.000 + 0.000 + 0.002 + 0.009 + 0.028 + 0.067 + 0.122 + 0.175$

$= 0.403$

(c) $\mu = np = 16(0.50) = 8$

The expected number of inmates serving time for drug dealing is 8.

7. $n = 10, p = 0.75$

(a) The probabilities can be obtained directed from Table 3 in Appendix II.

Number of Good Grapefruit

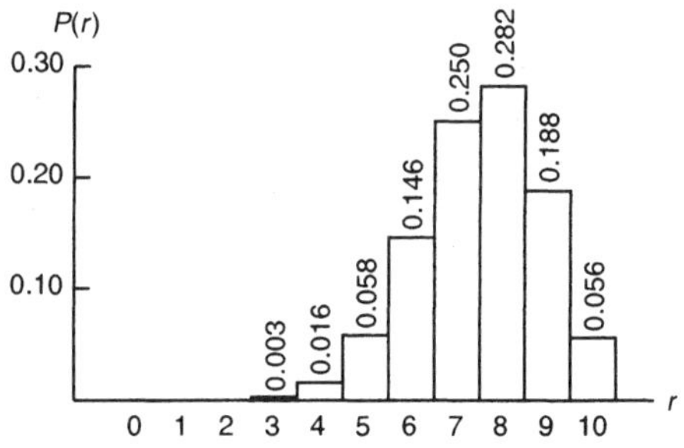

(b) No more than one bad is the same as at least nine good.

$P(r \geq 9) = P(9) + P(10)$

$= 0.188 + 0.056$

$= 0.244$

$P(r \geq 1) = P(1) + P(2) + P(3) + P(4) + P(5) + P(6) + P(7) + P(8) + P(9) + P(10)$

$= 0.000 + 0.000 + 0.003 + 0.016 + 0.058 + 0.146 + 0.250 + 0.282 + 0.188 + 0.056$

$= 0.999$

(c) $\mu = np = 10(0.75) = 7.5$

The expected number of good grapefruit in a sack is 7.5.

(d) $\sigma = \sqrt{npq} = \sqrt{10(0.75)(0.25)} \approx 1.37$

9. $p = 0.85$, $n = 12$

$$\begin{aligned} P(r \le 2) &= P(0) + P(1) + P(2) \\ &= 0.000 + 0.000 + 0.000 \\ &= 0.000 \quad \text{(to 3 digits)} \end{aligned}$$

The data seem to indicate that the percent favoring the increase in fees is less than 85%.

11. **(a)** Essay. Answer could include:

The Poisson distribution would be a good choice because coughs are a relatively rare occurrence. It is reasonable to assume that they are independent events, and the variable is the number of coughs in a fixed time interval.

(b) $\lambda = 11$ per 1 minute

From Table 4 in Appendix II,

$$\begin{aligned} P(r \le 3) &= P(0) + P(1) + P(2) + P(3) \\ &= 0.0000 + 0.0002 + 0.0010 + 0.0037 \\ &= 0.0049 \end{aligned}$$

(c) $\lambda = \dfrac{11}{60 \text{ sec}} \cdot \dfrac{0.5}{0.5} = \dfrac{5.5}{30 \text{ sec}}$; $\lambda = 5.5$ per 30 sec.

$$\begin{aligned} P(r \ge 3) &= 1 - P(0) - P(1) - P(2) \\ &= 1 - 0.0041 - 0.0225 - 0.0618 \\ &= 0.9116 \end{aligned}$$

13. The loan-default problem satisfies the conditions for a binomial experiment. Moreover, p is small, n is large, and $np < 10$. Use of the Poisson approximation to the binomial distribution is appropriate.

$$n = 300,\ p = \frac{1}{350} = 0.0029,\ \lambda = np = 300(0.0029) \approx 0.86 \approx 0.9$$

From Table 4 in Appendix II,

$$\begin{aligned} P(r \ge 2) &= 1 - P(0) - P(1) \\ &= 1 - 0.4066 - 0.3659 \\ &= 0.2275 \end{aligned}$$

15. **(a)** Use the geometric distribution with $p = 0.5$.

$$P(n = 2) = (0.5)(0.5) = (0.5)^2 = 0.25$$

As long as you toss the coin at least twice, it does not matter how many more times you toss it. To get the first head on the second toss, you must get a tail on the first and a head on the second.

(b) $P(4) = (0.5)(0.5)^3 = (0.5)^4 = 0.0625$

$$\begin{aligned} P(n > 4) &= 1 - P(1) - P(2) - P(3) - P(4) \\ &= 1 - 0.5 - 0.5^2 - 0.5^3 - 0.5^4 \\ &= 0.0625 \end{aligned}$$

Chapter 6 Normal Distributions

Section 6.1

1. (a) not normal; left skewed instead of symmetric

(b) not normal; curve touches and goes below x-axis instead of always being above the x-axis and being asymptotic to the x-axis in the tails

(c) not normal; not bell-shaped, not unimodal

(d) not normal; not a smooth curve

3. The mean is the x-value directly below the peak; in Figure 6-14, $\mu = 10$; in Figure 6-15, $\mu = 4$. Assuming the two figures are drawn on the same scale, Figure 6-14, being shorter and more spread out, has the larger standard deviation.

5. (a) 50%; the normal curve is symmetric about μ.

(b) 68%

(c) 99.7%

7. (a) $\mu = 65$, so 50% are taller than 65 in.

(b) $\mu = 65$, so 50% are shorter than 65 in.

(c) $\mu - \sigma = 65 - 2.5 = 62.5$ in. and $\mu + \sigma = 65 + 2.5 = 67.5$ in. so 68% of college women are between 62.5 in. and 67.5 in. tall.

(d) $\mu - 2\sigma = 65 - 2(2.5) = 65 - 5 = 60$ in. and $\mu - 2\sigma = 65 + 2(2.5) = 65 + 5 = 70$ in. so 95% of college women are between 60 in. and 70 in. tall.

9. (a) $\mu - \sigma = 1243 - 36 = 1207$ and $\mu + \sigma = 1243 + 36 = 1279$ so about 68% of the tree rings will date between 1207 and 1279 AD.

(b) $\mu - 2\sigma = 1243 - 2(36) = 1171$ and $\mu + 2\sigma = 1243 + 2(36) = 1315$ so about 95% of the tree rings will date between 1171 and 1315 AD.

(c) $\mu - 3\sigma = 1243 - 3(36) = 1135$ and $\mu + 3\sigma = 1243 + 3(36) = 1351$ so 99.7% (almost all) of the tree rings will date between 1135 and 1351 AD.

11. (a) $\mu - \sigma = 3.15 - 1.45 = 1.70$ and $\mu + \sigma = 3.15 + 1.45 = 4.60$ so 68% of the experimental group will have pain thresholds between 1.70 and 4.60 mA.

(b) $\mu - 2\sigma = 3.15 - 2(1.45) = 0.25$ and $\mu + 2\sigma = 3.15 + 2(1.45) = 6.05$ so 95% of the experimental group will have pain thresholds between 0.25 and 6.05 mA.

13. (a)

Tri-County Bank Monthly Loan Request—First Year
(thousands of dollars)

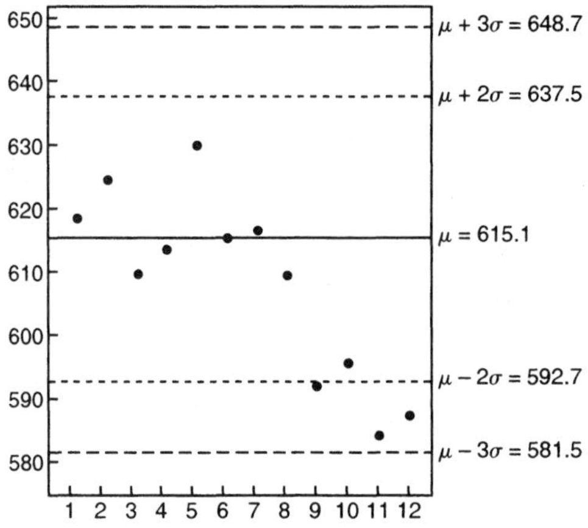

The economy would appear to be cooling off as evidenced by an overall downward trend. Out-of-control warning signal III is present: 2 of the last 3 consecutive points are below $\mu - 2\sigma = 592.7$.

(b)

Tri-County Bank Monthly Loan Request—Second Year (thousands of dollars)

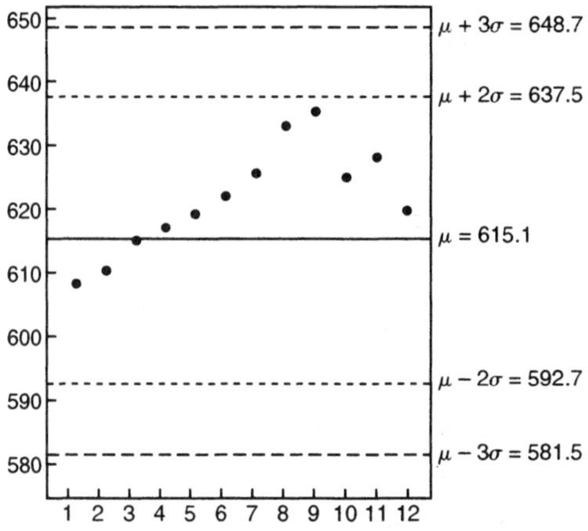

Here, it looks like the economy was heating up during months 1-9 and perhaps cooling off during months 10-12. Out-of-control warning signal II is present: there is a run of 9 consecutive points above $\mu = 615.1$.

15.

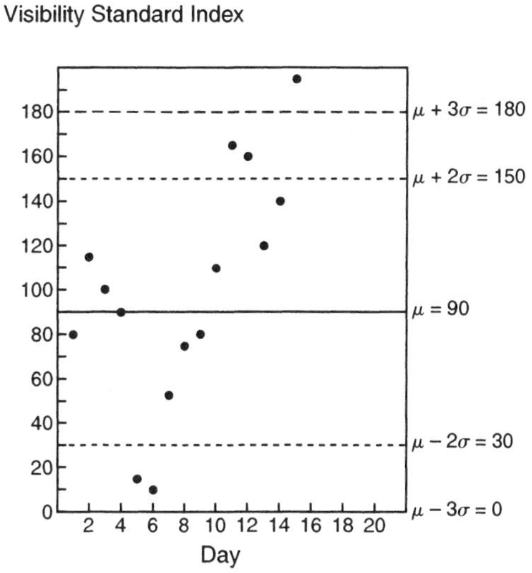

Out-of-control warning signals I and III are present. Day 15's VSI exceeds $\mu + 3\sigma$. Two of 3 consecutive points (days 10, 11, 12 or days 11, 12, 13) are about $\mu + 2\sigma = 150$, and 2 of 3 consecutive points (days 4, 5, 6 or days 5, 6, 7) are below $\mu - 2\sigma = 30$. Days 10-15 all show above average air pollution levels; days 11, 12, and 15 triggered out-of-control signals, indicating pollution abatement procedures should be in place.

Section 6.2

1. **(a)** z-scores > 0 indicate the student scored above the mean: Robert, Jan, and Linda

(b) z-scores = 0 indicates the student scored at the mean: Joel

(c) z-scores < 0 indicate the student scored below the mean: Juan and Susan

(d) $z = \frac{x-\mu}{\sigma}$ so $x = \mu + z\sigma$

In this case, if the student's score is x, $x = 150 + z(20)$.

Robert: $x = 150 + 1.10(20) = 172$

Joel: $x = 150 + 0(20) = 150$

Jan: $x = 150 + 1.70(20) = 184$

Juan: $x = 150 - 0.80(20) = 134$

Susan: $x = 150 - 2.00(20) = 110$

Linda: $x = 150 + 1.60(20) = 182$

3. Use $z = \frac{x-\mu}{\sigma}$. In this case, $z = \frac{x-73}{5}$.

(a) $53^\circ\text{F} < x < 93^\circ\text{F}$

$\frac{53-73}{5} < \frac{x-73}{5} < \frac{93-73}{5}$ Subtract $\mu = 73^\circ$F from each piece; divide result by $\sigma = 5^\circ$F.

$-\frac{20}{5} < z < \frac{20}{5}$

$-4.00 < z < 4.00$

(b) $x < 65°\text{F}$

$x - 73 < 65 - 73$ Subtract $\mu = 73°\text{F}$.

$\frac{x-73}{5} < \frac{65-73}{5}$ Divide both sides by $\sigma = 5°\text{F}$.

$z < -1.6$

(c) $78°\text{F} < x$

$\frac{78-73}{5} < \frac{x-73}{5}$ Subtract $\mu = 73°\text{F}$ from each side; divide by $\sigma = 5°\text{F}$.

$\frac{5}{5} < z$

$1.00 < z$ (or $z > 1.00$)

For (d)–(f): Since $z = \frac{x-73}{5}$, $x = 73 + 5z$.

(d) $1.75 < z$

$5(1.75) < 5z$ Multiply both sides by $\sigma = 5°\text{F}$.

$73 + 5(1.75) < 73 + 5z$ Add $\mu = 73°\text{F}$ to both sides.

$81.75°\text{F} < x$ (or $x > 81.75$)

(e) $z < -1.90$

$5z < 5(-1.90)$ Multiply both sides by $\sigma = 5°\text{F}$.

$73 + 5z < 73 + 5(-1.90)$ Add $\mu = 73°\text{F}$ to both sides.

$x < 63.5°\text{F}$

(f) $-1.80 < z < 1.65$

$5(-1.80) < 5z < 5(1.65)$ Multiply each part by $\sigma = 5°\text{F}$.

$73 + 5(-1.80) < 73 + 5z < 73 + 5(1.65)$ Add $\mu = 73°\text{F}$ to each part of the inequality.

$64°\text{F} < x < 81.25°\text{F}$

5. $z = \frac{x-\mu}{\sigma}$, here $z = \frac{x-4400}{620}$

(a) $3300 < x$

$3300 - 4400 < x - 4400$ Subtract $\mu = 4400$ deer.

$\frac{3300-4400}{620} < \frac{x-4400}{620}$ Divide by $\sigma = 620$ deer.

$-1.77 < z$

(b) $x < 5400$

$x - 4400 < 5400 - 4400$ Subtract $\mu = 4400$ deer.

$\frac{x-4400}{620} < \frac{5400-4400}{620}$ Divide by $\sigma = 620$ deer.

$z < 1.61$

(c) $3500 < x < 5300$

$3500 - 4400 < x - 4400 < 5300 - 4400$ Subtract $\mu = 4400$.

$\frac{3500\text{-}4400}{620} < \frac{x-4400}{620} < \frac{5300-4400}{620}$ Divide by $\sigma = 620$.

$-1.45 < z < 1.45$

For (d)–(f): Since $z = \dfrac{x - 4400}{620}$, $x = 4400 + 620z$ deer.

(d)

$$\begin{aligned} -1.12 &< z < 2.43 \\ 620(-1.12) &< 620z < 620(2.43) && \text{Multiply by } \sigma = 620. \\ 4400 + 620(-1.12) &< 4400 + 620z < 4400 + 620(2.43) && \text{Add } \mu = 4400 \text{ to each part.} \\ 3706 \text{ deer } &< x < 5907 \text{ deer (rounded)} \end{aligned}$$

(e)

$$\begin{aligned} z &< 1.96 \\ 620z &< 620(1.96) && \text{Multiply by } \sigma. \\ 4400 + 620z &< 4400 + 620(1.96) && \text{Add } \mu. \\ x &< 5615 \text{ deer} \end{aligned}$$

(f)

$$\begin{aligned} 2.58 &< z \\ 620(2.58) &< 620z && \text{Multiply by } \sigma. \\ 4400 + 620(2.58) &< 4400 + 620z && \text{Add } \mu. \\ 6000 \text{ deer } &< x \end{aligned}$$

(g) If $x = 2800$ deer, $z = \dfrac{2800 - 4400}{620} = -2.58$.

This is a small z-value, so 2800 deer is quite low for the fall deer population.

If $x = 6300$ deer, $z = \dfrac{6300 - 4400}{620} = 3.06$.

This is a very large z-value, so 6300 deer would be an unusually large fall population size.

7. $z = \dfrac{x - \mu}{\sigma}$, so in this case $z = \dfrac{x - 4.8}{0.3}$.

(a)

$$\begin{aligned} 4.5 &< x \\ \frac{4.5 - 4.8}{0.3} &< \frac{x - 4.8}{0.3} && \text{Subtract } \mu\text{; divide by } \sigma. \\ -1.00 &< z \end{aligned}$$

(b)

$$\begin{aligned} x &< 4.2 \\ \frac{x - 4.8}{0.3} &< \frac{4.2 - 4.8}{0.3} && \text{Subtract } \mu\text{; divide by } \sigma. \\ z &< -2.00 \end{aligned}$$

(c)

$$\begin{aligned} 4.0 &< x < 5.5 \\ \frac{4.0 - 4.8}{0.3} &< \frac{x - 4.8}{0.3} < \frac{5.5 - 4.8}{0.3} && \text{Subtract } \mu\text{; divide by } \sigma. \\ -2.67 &< z < 2.33 \end{aligned}$$

For (d)–(f): Since $z = \dfrac{x - 4.8}{0.3}$, $x = 4.8 + 0.3z$.

(d)

$$\begin{aligned} z &< -1.44 \\ 0.3z &< 0.3(-1.44) && \text{Multiply by } \sigma. \\ 4.8 + 0.3z &< 4.8 + 0.3(-1.44) && \text{Add } \mu. \\ x &< 4.4 \end{aligned}$$

(e)

$$\begin{aligned} 1.28 &< z \\ 0.3(1.28) &< 0.3z \quad \text{Multiply by } \sigma. \\ 4.8+0.3(1.28) &< 4.8+0.3z \quad \text{Add } \mu. \\ 5.2 &< x \end{aligned}$$

(f)

$$\begin{aligned} -2.25 < z &< -1.00 \\ 0.3(-2.25) < 0.3z &< 0.3(-1.00) \quad \text{Multiply by } \sigma. \\ 4.8+0.3(-2.25) < 4.8+0.3z &< 4.8+0.3(-1.00) \quad \text{Add } \mu. \\ 4.1 < x &< 4.5 \end{aligned}$$

(g) If the RBC was 5.9 or higher, that would be an unusually high red blood cell count.

$$x \geq 5.9$$

$$\frac{x-4.8}{0.3} \geq \frac{5.9-4.8}{0.3}$$

$z \geq 3.67$ (a very large z-value)

For problems 9–47, refer to the following sketch patterns for guidance in calculations

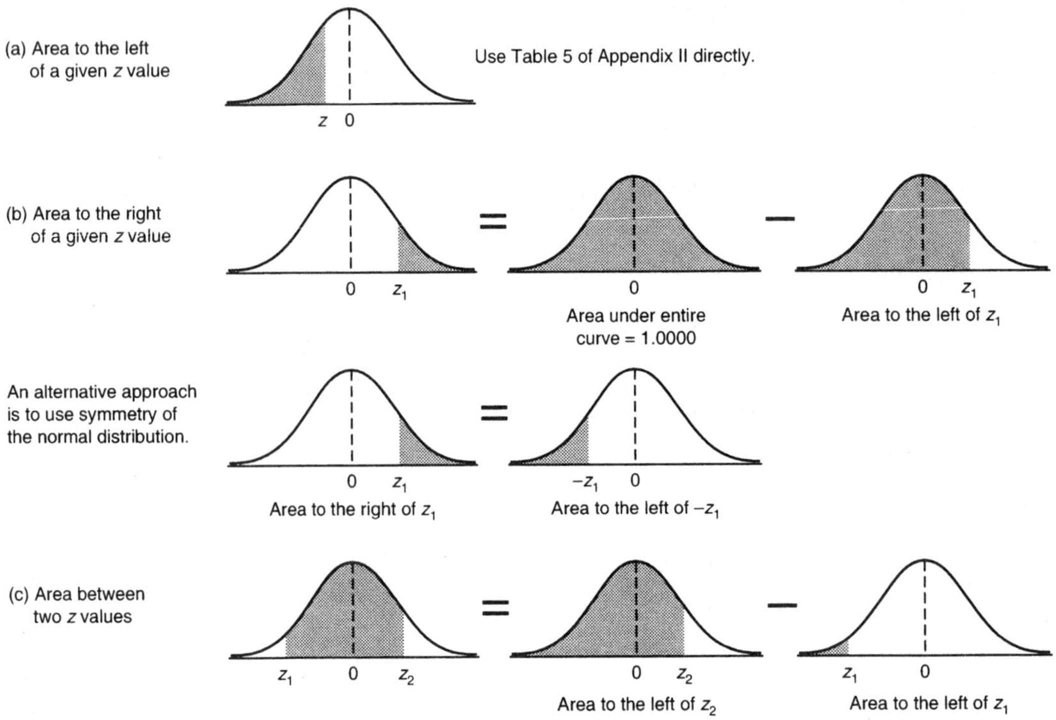

Using the left-tail style standard normal distribution table (see figures above)

(a) For areas to the *left* of a specified z value, use the table entry directly.

(b) For areas to the *right* of a specified z value, look up the table entry for z and subtract the table value from 1. (This is the complementary event rule as applied to area as probability.)

OR: Use the fact that the normal curve is symmetric about the mean, 0. The area in the right tail above a z-value is the same as the area in the left tail below the value of $-z$. To find the area to the right of z, look up the table value for $-z$.

(c) For areas *between* two z-values, z_1 and z_2, where $z_1 < z_2$, subtract the tabled value for z_1, from the tabled value for z_2.

These sketches and rules for finding the area for probability from the standard normal table apply for *any* z: $-\infty < z < +\infty$.

Student sketches should resemble those indicated with negative z-values to left of 0 and positive z-values to the right of zero.

9. Refer to figure (b).
The area to the right of $z = 0$ is 1 – area to left of $z = 0$, or $1 - 0.5000 = 0.5000$.

11. Refer to figure (a).
The area to the left of $z = -1.32$ is 0.0934.

13. Refer to figure (a).
The area to left of $z = 0.45$ is 0.6736.

15. Refer to figure (b).
The area to right of $z = 1.52$ is $1 - 0.9357 = 0.0643$.

17. Refer to figure (b).
The area to right of $z = -1.22$ is $1 - 0.1112 = 0.8888$.

19. Refer to figure (c).
The area between $z = 0$ and $z = 3.18$ is $0.9993 - 0.5000 = 0.4993$.

21. Refer to figure (c).
The area between $z = 0$ and $z = -2.01$ is $0.5000 - 0.0222 = 0.4778$.

23. Refer to figure (c).
The area between $z = -2.18$ and $z = 1.34$ is $0.9099 - 0.0146 = 0.8953$.

25. Refer to figure (c).
The area between $z = 0.32$ and $z = 1.92$ is $0.9726 - 0.6255 = 0.3471$.

27. Refer to figure (c).
The area between $z = -2.42$ and $z = -1.77$ is $0.0384 - 0.0078 = 0.0306$.

29. Refer to figure (a).
$P(z \leq 0) = 0.5000$

31. Refer to figure (a).
$P(z \leq -0.13) = 0.4483$ (direct read)

33. Refer to figure (a).
$P(z \leq 1.20) = 0.8849$

35. Refer to figure (b).
$P(z \geq 1.35) = 1 - P(z < 1.35) = 1 - 0.9115 = 0.0885$

37. Refer to figure (b).
$P(x \geq -1.20) = 1 - P(z < -1.20) = 1 - 0.1151 = 0.8849$

39. Refer to figure (c).
$P(-1.20 \le z \le 2.64) = P(z \le 2.64) - P(z < -1.20) = 0.9959 - 0.1151 = 0.8808$

41. Refer to figure (c).
$P(-2.18 \le z \le -0.42) = P(z \le -0.42) - P(z < -2.18) = 0.3372 - 0.0146 = 0.3226$

43. Refer to figure (c).
$P(0 \le z \le 1.62) = P(z \le 1.62) - P(z < 0) = 0.9474 - 0.5000 = 0.4474$

45. Refer to figure (c).
$P(-0.82 \le z \le 0) = P(z \le 0) - P(z < -0.82) = 0.5000 - 0.2061 = 0.2939$

47. Refer to figure (c).
$P(-0.45 \le z \le 2.73) = P(z \le 2.73) - P(z < -0.45) = 0.9968 - 0.3264 = 0.6704$

Section 6.3

1. We are given $\mu = 4$ and $\sigma = 2$. Since $z = \frac{x-\mu}{\sigma}$, we have $z = \frac{x-4}{2}$.

$P(3 \le x \le 6)$
$= P(3-4 \le x-4 \le 6-4)$ Subtract $\mu = 4$ from each part of the inequality.
$= P\left(\frac{3-4}{2} \le \frac{x-4}{2} \le \frac{6-4}{2}\right)$ Divide each part by $\sigma = 2$.
$= P\left(-\frac{1}{2} \le z \le \frac{2}{2}\right)$
$= P(-0.5 \le z \le 1)$
$= P(z \le 1) - P(z < -0.5)$ Refer to sketch (c) in the solutions for Section 6.2.
$= 0.8413 - 0.3085$
$= 0.5328$

3. We are given $\mu = 40$ and $\sigma = 15$. Since $z = \frac{x-\mu}{\sigma}$, we have $z = \frac{x-40}{15}$.

$P(50 \le x \le 70)$
$= P(50-40 \le x-40 \le 70-40)$ Subtract $\mu = 40$.
$= P\left(\frac{50-40}{15} \le \frac{x-40}{15} \le \frac{70-40}{15}\right)$ Divide by $\sigma = 15$.
$= P(0.67 \le z \le 2)$
$= P(z \le 2) - P(z < 0.67)$
$= 0.9772 - 0.7486 = 0.2286$

5. We are given $\mu = 15$ and $\sigma = 3.2$. Since $z = \dfrac{x-\mu}{\sigma}$, we have $z = \dfrac{x-15}{3.2}$.

$$\begin{aligned}
&P(8 \le x \le 12)\\
&= P(8-15 \le x-15 \le 12-15) && \text{Subtract } \mu = 15.\\
&= P\left(\frac{8-15}{3.2} \le \frac{x-15}{3.2} \le \frac{12-15}{3.2}\right) && \text{Divide by } \sigma = 3.2.\\
&= P(-2.19 \le z \le -0.94)\\
&= P(z \le -0.94) - P(z < -2.19)\\
&= 0.1736 - 0.0143 = 0.1593
\end{aligned}$$

7. We are given $\mu = 20$ and $\sigma = 3.4$. Since $z = \dfrac{x-\mu}{\sigma}$, we have $z = \dfrac{x-20}{3.4}$.

$$\begin{aligned}
&P(x \ge 30)\\
&= P(x-20 \ge 30-20) && \text{Subtract } \mu = 20.\\
&= P\left(\frac{x-20}{3.4} \ge \frac{30-20}{3.4}\right) && \text{Divide by } \sigma = 3.4.\\
&= P(z \ge 2.94)\\
&= 1 - P(z < 2.94) && \text{Refer to sketch (b) in Section 6.2.}\\
&= 1 - 0.9984 = 0.0016
\end{aligned}$$

9. We are given $\mu = 100$ and $\sigma = 15$. Since $z = \dfrac{x-\mu}{\sigma}$, we have $z = \dfrac{x-100}{15}$.

$$\begin{aligned}
&P(x \ge 90)\\
&= P\left(\frac{x-100}{15} \ge \frac{90-100}{15}\right) && \text{Subtract } \mu\text{; divide by } \sigma.\\
&= P(z \ge -0.67)\\
&= 1 - P(z < -0.67) = 1 - 0.2514 = 0.7486
\end{aligned}$$

For problems 11–19, refer to the following sketch patterns for guidance in calculation.

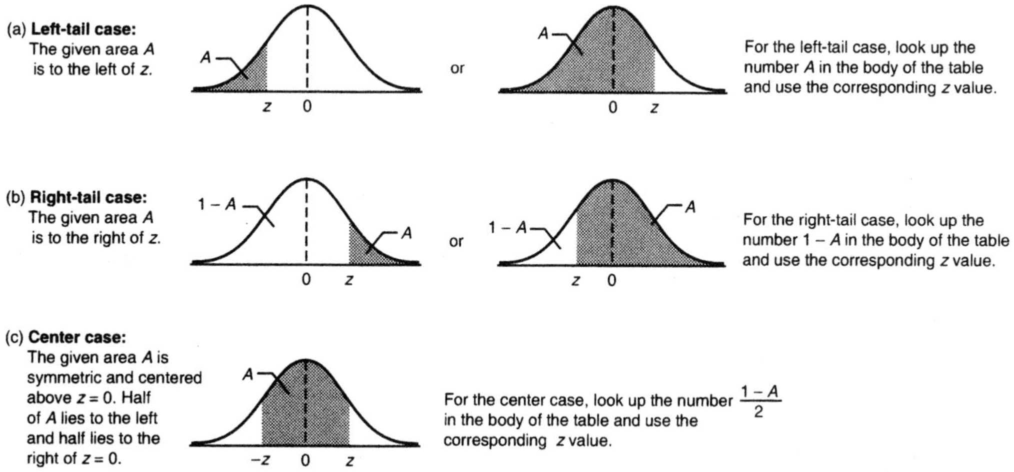

(a) Left-tail case:
The given area A is to the left of z.
A
z 0
or
A
0 z
For the left-tail case, look up the number A in the body of the table and use the corresponding z value.
(b) Right-tail case:
The given area A is to the right of z.
1 − A
A
0 z
or
1 − A
A
z 0
For the right-tail case, look up the number 1 − A in the body of the table and use the corresponding z value.
(c) Center case:
The given area A is symmetric and centered above z = 0. Half of A lies to the left and half lies to the right of z = 0.
A
−z 0 z
For the center case, look up the number (1 − A)/2 in the body of the table and use the corresponding z value.

Student sketches should resemble the figures above, with negative z-values to the left of zero and positive z-values to the right of zero, and A written as a decimal.

11. Refer to figure (a).
Find z so that the area A to the left of z is 6% = 0.06. Since A = 0.06 is less than 0.5000, look for a negative z value. A to left of –1.55 is 0.0606 and A to left of –1.56 is 0.0594. Since 0.06 is in the middle of 0.0606 and 0.0594, for our z-value we will use the average of –1.55 and –1.56:

$$\frac{-1.55+(-1.56)}{2} = -1.555.$$

13. Refer to figure (a).
Find z so that the area A to the left of z is 55% = 0.55. Since A = 0.55 > 0.5000, look for a positive z-value. The area to the left of 0.13 is 0.5517, so z = 0.13.

15. Refer to figure (b).
Find z so that the area A to the right of z is 8% = 0.08. Since A to the right of z is 0.08, 1 – A = 1 – 0.08 = 0.92 is to the left of z-value. The area to the left of 1.41 is 0.9207.

17. Refer to figure (b).
Find z so that the area A to the right of z is 82% = 0.82. Since A to the right of z, 1 – A = 1 – 0.82 = 0.18 is to the left of z. Since 1 – A = 0.18 < 0.5000, look for a negative z value. The area to the left of z = –0.92 is 0.1788.

19. Refer to figure (c).
Find z such that the area A between $-z$ and z is 98% = 0.98. Since A is between $-z$ and z, 1 – A = 1 – 0.98 = 0.02 lies in the tails, and since we need $\pm z$, half of 1 – A lies in each tail. The area to the left of $-z$ is $\frac{1-A}{2} = \frac{0.02}{2} = 0.01$. The area to the left of –2.33 is 0.0099. Thus $-z$ = –2.33 and z = 2.33.

21. x is approximately normal with $\mu = 85$ and $\sigma = 25$. Since $z = \frac{x-\mu}{\sigma}$, we have $z = \frac{x-85}{25}$.

(a) $P(x > 60)$

$$= P\left(\frac{x-85}{25} > \frac{60-85}{25}\right) = P(z > -1)$$
$$= 1 - P(z \le -1) = 1 - 0.1587 = 0.8413$$

(b) $P(x < 110) = P\left(\frac{x-85}{25} < \frac{110-85}{25}\right) = P(z < 1) = 0.8413.$

(c) $P(60 < x < 110)$

$$= P(-1 < z < 1) \quad \text{using (a) and (b)}$$
$$= P(z < 1) - P(z \le -1) = 0.8413 - 0.1587 = 0.6826$$

(i.e., approximately 68% of the blood glucose measurements lie within $\mu \pm \sigma$)

(d) $P(x > 140)$

$$= P\left(\frac{x-85}{25} > \frac{140-85}{25}\right) = P(z > 2.2)$$
$$= 1 - P(z \le 2.2) = 1 - 0.9861 = 0.0139$$

23. SAT scores, x, are normal with $\mu_x = 500$ and $\sigma_x = 100$. Since $z = \frac{x-\mu_x}{\sigma_x}$, we have $x = \frac{x-500}{100}$.

(a) $P(x > 675) = P\left(\frac{x-500}{100} > \frac{675-500}{100}\right) = P(z > 1.75) = 1 - P(z \le 1.75) = 1 - 0.9599 = 0.0401$

(b) $P(x < 450) = P\left(\frac{x-500}{100} < \frac{450-500}{100}\right) = P(z < -0.5) = 0.3085$

(c) $P(450 \le x \le 675)$
$= P(-0.5 \le z \le 1.75)$ using (a), (b)
$= P(z \le 1.75) - P(z < -0.5) = 0.9599 - 0.3085$
$= 0.6514$ using work in (a), and (b)

ACT scores, y, are normal with $\mu_y = 18$ and $\sigma_y = 6$. Since $z = \frac{y - \mu_y}{\sigma_y}$, we have $z = \frac{y-18}{6}$.

(d) $P(y > 28) = P\left(\frac{y-18}{6} > \frac{28-18}{6}\right) = P(z > 1.67) = 1 - P(z \le 1.67) = 1 - 0.9525 = 0.0475$

(e) $P(y > 12) = P\left(\frac{y-18}{6} > \frac{12-18}{6}\right) = P(z > -1) = 1 - P(z \le -1) = 1 - 0.1587 = 0.8413$

(f) $P(12 \le y \le 28)$
$= P(-1 \le z \le 1.67)$ using (a), (b)
$= P(z \le 1.67) - P(z < -1) = 0.9525 - 0.1587 = 0.7938$

25. Pot shard thickness, x, is approximately normally distributed with $\mu = 5.1$ and $\sigma = 0.9$ millimeters.

(a) $P(x < 3.0) = P\left(\frac{x-5.1}{0.9} < \frac{3.0-5.1}{0.9}\right) = P(z < -2.33) = 0.0099$

(b) $P(x > 7.0) = P\left(\frac{x-5.1}{0.9} > \frac{7.0-5.1}{0.9}\right) = P(z > 2.11) = 1 - P(z \le 2.11) = 1 - 0.9826 = 0.0174$

(c) $P(3.0 \le x \le 7.0)$
$= P(-2.33 \le z \le 2.11)$ using (a), (b)
$= P(z \le 2.11) - P(z < -2.33) = 0.9826 - 0.0099$
$= 0.9727$

27. Fuel consumption, x, is approximately normal with $\mu = 3213$ and $\sigma = 180$ gallons per hour.

(a) $P(3000 \le x \le 3500)$
$= P\left(\frac{3000-3213}{180} \le \frac{x-3213}{180} \le \frac{3500-3213}{180}\right)$
$= P(-1.18 \le z \le 1.59) = P(z \le 1.59) - P(z < -1.18)$
$= 0.9441 - 0.1190 = 0.8251$

(b) $P(x < 3000) = P(z < -1.18) = 0.1190$ using (a)

(c) $P(x > 3500) = P(z > 1.59) = 1 - P(z \le 1.59) = 1 - 0.9441 = 0.0559$ using (a)

29. Lifetime, x, is normally distributed with $\mu = 45$ and $\sigma = 8$ months.

(a) $P(x \le 36) = P\left(\frac{x-45}{8} \le \frac{36-45}{8}\right) = P(z \le -1.125) \approx P(z \le -1.13) = 0.1292$
The company will have to replace approximately 13% of its batteries.

(b) Find x_0 such that $P(x \le x_0) = 10\% = 0.10$. First, find z_0 such that $P(z \le z_0) = 0.10$.
$P(z \le -1.28) = 0.1003$, so $z_0 = -1.28$. Then $x_0 = \mu + z_0\sigma = 45 + (-1.28)(8) = 34.76 \approx 35$.
The company should guarantee the batteries for 35 months.

31. Age at replacement, x, is approximately normal with $\mu = 8$ and range = 6 years.

(a) The empirical rule says that about 95% of the data are between $\mu - 2\sigma$ and $\mu + 2\sigma$, or about 95% of the data are in a $(\mu + 2\sigma) - (\mu - 2\sigma) = 4\sigma$ range (centered around μ). Thus, the range $\approx 4\sigma$, or $\sigma \approx$ range/4. Here, we can approximate σ by 6/4 = 1.5 years.

(b) $P(x > 5) = P\left(z > \frac{5-8}{1.5}\right) = P(z > -2)$ using the estimate of σ from (a)
$= 1 - P(z \le -2) = 1 - 0.0228 = 0.9772$

(c) $P(x < 10) = P\left(z < \frac{10-8}{1.5}\right) = P(z < 1.33) = 0.9082$

(d) Find x_0 so that $P(x \le x_0) = 10\% = 0.10$. First, find z_0 such that $P(z \le z_0) = 0.10$.
$P(z \le -1.28) = 0.1003$, so $z_0 = -1.28$. Then $x_0 = \mu + z_0\sigma = 8 + (-1.28)(1.5) = 6.08$.
The company should guarantee their TVs for about 6.1 years.

33. Resting heart rate, x, is approximately normal with $\mu = 46$ and (95%) range from 22 to 70 bpm.

(a) From Problem 31(a), range $\approx 4\sigma$, or $\sigma \approx$ range/4. Here range = 70 – 22 = 48, so $\sigma \approx 48/4 = 12$ bpm.

(b) $P(x < 25) = P\left(z < \frac{25-46}{12}\right) = P(z < -1.75) = 0.0401$

(c) $P(x > 60) = P\left(z > \frac{60-46}{12}\right) = P(z > 1.17) = 1 - P(z \le 1.17) = 1 - 0.8790 = 0.1210$

(d) $P(25 \le x \le 60) = P(-1.75 \le z \le 1.17)$ using (b), (c)
$= P(z \le 1.17) - P(z < -1.75)$
$= 0.8790 - 0.0401$
$= 0.8389$

(e) Find x_0 such that $P(x > x_0) = 10\% = 0.10$. First, find z_0 such that $P(z > z_0) = 0.10$.
$P(z \le z_0) = 1 - 0.10 = 0.90$
$P(z \le 1.28) = 0.8997 \approx 0.90$, so let $z_0 = 1.28$.
When $x_0 = \mu + z_0\sigma = 46 + 1.28(12) = 61.36$, so horses with resting rates of 61 bpm or more may need treatment.

35. Life expectancy x is normal with $\mu = 90$ and $\sigma = 3.7$ months.

(a) The insurance company wants 99% of the microchips to last longer than x_0. Saying this another way: the insurance company wants to pay the \$50 million at most 1% of the time. So, find x_0 such that $P(x \le x_0) = 1\% = 0.01$. First, find z_0 such that $P(z \le z_0) = 0.01$. $P(z \le -2.33) = 0.0099 \approx 0.01$, so let $z_0 = -2.33$. Since $z_0 = \frac{x_0 - \mu}{\sigma}$, $x_0 = \mu + z_0\sigma = 90 + (-2.33)(3.7) = 81.379 \approx 81$ months.

(b) $P(x \le 84) = P\left(z \le \frac{84-90}{3.7}\right) = P(z \le -1.62) = 0.0526 \approx 5\%$.

(c) The "expected loss" is 5.26% [from (b)] of the \$50 million, or 0.0526(50,000,000) = \$2,630,000.

(d) Profit is the difference between the amount of money taken in (here, \$3 million), and the amount paid out (here, \$2.63 million, from (c)). So the company expects to profit 3,000,000 – 2,630,000 = \$370,000.

37. Arrival times are normal with $\mu = 3$ hours 48 minutes and $\sigma = 52$ minutes after the doors open. Convert μ to minutes: $(3 \times 60) + 48 = 228$ minutes.

(a) Find x_0 such that $P(x \le x_0) = 90\% = 0.90$. First, find z_0 such that $P(z \le z_0) = 0.90$.

$P(z \le 1.28) = 0.8997 \approx 0.90$, so let $z_0 = 1.28$. Since $z_0 = \frac{x_0 - \mu}{\sigma}$, $x_0 = \mu + z_0\sigma = 228 + (1.28)(52)$

$= 294.56$ minutes, or $294.56/60 = 4.9093 \approx 4.9$ hours after the doors open.

(b) Find x_0 such that $P(x \le x_0) = 15\% = 0.15$. First, find z_0 such that $P(z \le z_0) = 0.15$.

$P(z \le -1.04) = 0.1492 \approx 0.15$, so let $z_0 = -1.04$. Then $x_0 = \mu + z_0\sigma = 228 + (-1.04)(52)$

$= 173.92$ minutes, or $173/60 = 2.899 \approx 2.9$ hours after the doors open.

(c) Answers vary. Most people have Saturday off, so many may come early in the day. Most people work Friday, so most people would probably come after 5 P.M. There is no reason to think weekday and weekend arrival times would have the same distribution.

39. Waiting time, x, is approximately normal with $\mu = 18$ and $\sigma = 4$ minutes.

(a) Let A be the event that $x > 20$, and B be the event that $x > 15$. We want to find $P(A, \textit{given } B)$.

Recall $P(A, \textit{given } B) = \dfrac{P(A \textit{ and } B)}{P(B)}$.

$$P(A \textit{ and } B) = P(x > 20 \textit{ and } x > 15) = P(x > 20)$$

Use a number line to find where both events occur simultaneously.

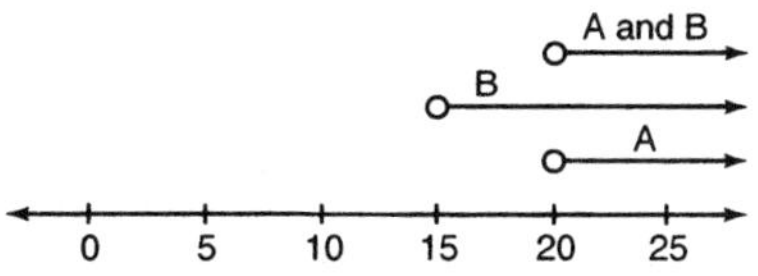

The number 20 is not included in "both A and B" because A says x is strictly greater than 20. The intervals $(15, \infty)$ and $(20, \infty)$ intersect at $(20, \infty)$.

$$P(x > 20) = P\left(z > \frac{20-18}{4}\right) = P(z > 0.5) = 1 - P(z \le 0.5) = 1 - 0.6915 = 0.3085$$

$$P(x > 15) = P\left(z > \frac{15-18}{4}\right) = P(z > -0.75) = 1 - P(z \le -0.75) = 1 - 0.2266 = 0.7734$$

$$\begin{aligned} P(x > 20, \textit{ given } x > 15) &= \frac{P(x > 20 \textit{ and } x > 15)}{P(x > 15)} \\ &= \frac{P(x > 20)}{P(x > 15)} \\ &= \frac{0.3085}{0.7734} \\ &= 0.3989 \end{aligned}$$

(b) $$\begin{aligned}P(x>25,\ given\ x>18) &= \frac{P(x>25\ and\ x>18)}{P(x>18)}\\ &= \frac{P(x>25)}{P(x>18)}\\ &= \frac{P\left(z>\frac{25-18}{4}\right)}{P\left(z>\frac{18-18}{4}\right)}\\ &= \frac{P(z>1.75)}{P(z>0)}\\ &= \frac{1-P(z\le 1.75)}{1-P(z\le 0)}\\ &= \frac{1-0.9599}{1-0.5000}\\ &= \frac{0.0401}{0.5}\\ &= 0.0802\end{aligned}$$

(c) Let event $A = x > 20$ and event $B = x > 15$ in the formula.

Section 6.4

Answers may vary slightly due to rounding.

1. Previously, $p = 88\% = 0.88$; now, $p = 9\% = 0.09$; $n = 200$; $r = 50$

Let a success be defined as a child with a high blood-lead level.

(a) $P(r \ge 50) = P(50 \le r) = P(49.5 \le x)$

$np = 200(0.88) = 176$; $nq = n(1-p) = 200(0.12) = 24$

Since both np and nq are greater than 5, we will use the normal approximation to the binomial with $\mu = np = 176$ and $\sigma = \sqrt{npq} = \sqrt{200(0.88)(0.12)} = \sqrt{21.12} = 4.60$.

So, $P(r \ge 50) = P(49.5 \le x) = P\left(\frac{49.5-176}{4.6} \le z\right) = P(-27.5 \le z)$.

Almost every z value will be greater than or equal to –27.5, so this probability is approximately 1. It is almost certain that 50 or more children a decade ago had high blood-lead levels.

(b) $P(r \ge 50) = P(50 \le r) = P(49.5 \le x)$

In this case, $np = 200(0.09) = 18$ and $nq = 200(0.91) = 182$, so both are greater than 5. Use the normal approximation with $\mu = np = 18$ and $\sigma = \sqrt{npq} = \sqrt{200(0.09)(0.91)} = \sqrt{16.38} = 4.05$.

So $P(49.5 \le x) = P\left(\frac{49.5-18}{4.05} \le z\right) = P(7.78 \le z)$.

Almost no z values will be larger than 7.78, so this probability is approximately 0. Today, it is almost impossible that a sample of 200 children would include at least 50 with high blood-lead levels.

3. We are given $n = 125$ and $p = 17\% = 0.17$. Let a success be defined as the police receiving enough information to locate and arrest a fugitive within 1 week.

(a) $P(r \geq 15) = P(15 \leq r) = P(14.5 \leq x)$. 15 is a left endpoint. $np = 125(0.17) = 21.25$ and $nq = 125(1-0.17) = 125(0.83) = 103.75$, which are both greater than 5, so we can use the normal approximation with $\mu = np = 21.25$ and $\sigma = \sqrt{npq} = \sqrt{125(0.17)(0.83)} = \sqrt{17.6375} = 4.20$. So

$$P(14.5 \leq x) = P\left(\frac{14.5-21.25}{4.2} \leq z\right) = P(-1.61 \leq z) = P(z \geq -1.61) = 1-P(z < -1.61) = 1-0.0537 = 0.9463.$$

(b) $P(r \geq 28) = P(28 \leq r)$ 28 is a left endpoint.

$$\begin{aligned} &= P(27.5 \leq x) \\ &= P\left(\frac{27.5-21.25}{4.2} \leq z\right) \\ &= P(1.49 \leq z) \\ &= P(z \geq 1.49) \\ &= 1-P(z < 1.49) \\ &= 1-0.9319 \\ &= 0.0681 \end{aligned}$$

(c) Remember, r "between" a and b is $a \leq r \leq b$.

$P(15 \leq r \leq 28) = P(14.5 \leq x \leq 28.5)$ 15 is a left endpoint and 28 is a right endpoint.

$$\begin{aligned} &= P\left(\frac{14.5-21.25}{4.2} \leq z \leq \frac{28.5-21.25}{4.2}\right) \\ &= P(-1.61 \leq z \leq 1.73) \\ &= P(z \leq 1.73) - P(z < -1.61) \\ &= 0.9582-0.0537 \\ &= 0.9045 \end{aligned}$$

(d) $n = 125, p = 0.17, q = 1-p = 1-0.17 = 0.83$.

np and nq are both greater than 5, so the normal approximation is appropriate.

5. We are given $n = 753$ and $p = 3.5\% = 0.035$; $q = 1 - p = 1 - 0.035 = 0.965$.
Let a success be a person living past age 90.

(a) $P(r \geq 15) = P(15 \leq r) = P(14.5 \leq x)$ 15 is a left endpoint.

Here, $np = 753(0.035) = 26.355$, and $nq = 753(0.965) = 726.645$, both of which are greater than 5; the normal approximation is appropriate, using $\mu = np = 26.355$ and

$\sigma = \sqrt{npq} = \sqrt{753(0.035)(0.965)} = \sqrt{25.4326} = 5.0431$.

$$\begin{aligned} P(14.5 \leq x) &= P\left(\frac{14.5-26.355}{5.0431} \leq z\right) \\ &= P(-2.35 \leq z) \\ &= P(z \geq -2.35) \\ &= 1-P(z < -2.35) \\ &= 1-0.0094 \\ &= 0.9906 \end{aligned}$$

(b) $P(r \geq 30) = P(30 \leq r)$
$= P(29.5 \leq x)$
$= P\left(\frac{29.5 - 26.355}{5.0431} \leq z\right)$
$= P(0.62 \leq z)$
$= P(z \geq 0.62)$
$= 1 - P(z < 0.62)$
$= 1 - 0.7324$
$= 0.2676$

(c) $P(25 \leq r \leq 35) = P(24.5 \leq x \leq 35.5)$
$= P\left(\frac{24.5 - 26.355}{5.0431} \leq z \leq \frac{35.5 - 26.355}{5.0431}\right)$
$= P(-0.37 \leq z \leq 1.81)$
$= P(z \leq 1.81) - P(z < -0.37)$
$= 0.9649 - 0.3557$
$= 0.6092$

(d) $P(r > 40) = P(r \geq 41)$
$= P(41 \leq r)$
$= P(40.5 \leq x)$
$= P\left(\frac{40.5 - 26.355}{5.0431} \leq z\right)$
$= P(2.80 \leq z)$
$= P(z \geq 2.80)$
$= 1 - P(z < 2.80)$
$= 1 - 0.9974$
$= 0.0026$

7. $n = 66$; $p = 80\% = 0.80$; $q = 1 - p = 1 - 0.80 = 0.20$
A success is when a new product fails within 2 years.

(a) $P(r \geq 47) = P(47 \leq r) = P(46.5 \leq x)$
$np = 66(0.80) = 52.8$, and $nq = 66(0.20) = 13.3$. Both exceed 5, so the normal approximation with $\mu = np = 52.8$ and $\sigma = \sqrt{npq} = \sqrt{66(0.8)(0.2)} = \sqrt{10.56} = 3.2496$ is appropriate.

$P(46.5 \leq x) = P\left(\frac{46.5 - 52.8}{3.2496} \leq z\right)$
$= P(-1.94 \leq z)$
$= P(z \geq -1.94)$
$= 1 - P(z < -1.94)$
$= 1 - 0.0262$
$= 0.9738$

(b) $P(r \leq 58) = P(x \leq 58.5) = P\left(z \leq \frac{58.5 - 52.8}{3.2496}\right) = P(z \leq 1.75) = 0.9599$

For (c) and (d), note we are interested now in products succeeding, so a success is redefined to be a new product staying on the market for 2 years. Here, $n = 66$, p is now 0.20 with q is now 0.80 (p and q above have been switched. Now $np = 13.2$ and $nq = 52.8$, $\mu = 13.2$, and σ stays equal to 3.2496.

(c) $P(r \geq 15) = P(15 \leq r)$
$= P(14.5 \leq x)$
$= P\left(\frac{14.5-13.2}{3.2496} \leq z\right)$
$= P(0.40 \leq z)$
$= P(z \geq 0.40)$
$= 1 - P(z < 0.40)$
$= 1 - 0.6554$
$= 0.3446$

(d) $P(r < 10) = P(r \leq 9) = P(x \leq 9.5) = P\left(z \leq \frac{9.5-13.2}{3.2496}\right) = P(z \leq -1.14) = 0.1271$

9. $n = 430$; $p = 70\% = 0.70$; $q = 1 - p = 1 - 0.70 = 0.30$
A success is finding the address or lost acquaintances.

(a) $P(r > 280) = P(r \geq 281) = P(281 \leq r) = P(280.5 \leq x)$
$np = 430(0.7) = 301$ and $nq = 430(0.3) = 129$
Since both np and nq are greater than 5, the normal approximation with $\mu = np = 301$ and $\sigma = \sqrt{npq} = \sqrt{430(0.7)(0.3)} = \sqrt{90.3} = 9.5026$ is appropriate.

$$P(280.5 \leq x) = P\left(\frac{280.5-301}{9.5026} \leq z\right) = P(-2.16 \leq z) = 1 - P(z < -2.16) = 1 - 0.0154 = 0.9846$$

(b) $P(r \geq 320) = P(320 \leq r)$
$= P(319.5 \leq x)$
$= P\left(\frac{319.5-301}{9.5026} \leq z\right)$
$= P(1.95 \leq z)$
$= 1 - P(z > 1.95)$
$= 1 - 0.9744$
$= 0.0256$

(c) $P(280 \leq r \leq 320) = P(279.5 \leq x \leq 320.5)$
$= P\left(\frac{279.5-301}{9.5026} \leq z \leq \frac{320.5-301}{9.5026}\right)$
$= P(-2.26 \leq z \leq 2.05)$
$= P(z \leq 2.05) - P(z < -2.26)$
$= 0.9798 - 0.0119$
$= 0.9679$

(d) $n = 430$, $p = 0.7$, $q = 0.3$
Both np and nq are greater than 5 so the normal approximation is appropriate, See (a).

11. $n = 850$, $p = 57\% = 0.57$, $q = 0.43$
Success = pass Ohio bar exam

(a) $P(r \ge 540) = P(540 \le r) = P(539.5 \le x)$

$np = 484.5$, $nq = 365.5$, $\mu = np = 484.5$, $\sigma = \sqrt{npq} = \sqrt{208.335} = 14.4338$

Since both np and nq are greater than 5, use normal approximation with μ and σ as above.

$$P(539.5 \le x) = P\left(\frac{539.5 - 484.5}{14.4338} \le z\right) = P(3.81 \le z) \approx 0 \text{ (to three digits)}$$

(b) $P(r \le 500) = P(x \le 500.5) = P\left(z \le \frac{500.5 - 484.5}{14.4338}\right) = P(z \le 1.11) = 0.8665$

(c)
$$\begin{aligned} P(485 \le r \le 525) &= P(484.5 \le x \le 525.5) \\ &= P\left(0 \le z \le \frac{525.5 - 484.5}{14.4338}\right) \\ &= P(0 \le z \le 2.84) \\ &= P(z \le 2.84) - P(z < 0) \\ &= 0.9977 - 0.5 \\ &= 0.4977 \end{aligned}$$

13. $n = 317$, $P(\text{buy, given sampled}) = 37\% = 0.37$, $P(\text{sampled}) = 60\% = 0.60 = p$ so $q = 0.40$

(a) $P(180 < r) = P(181 \le r) = P(180.5 \le x)$

$np = 190.2$, $nq = 126.8$, $\sigma = \sqrt{npq} = \sqrt{76.08} = 8.7224$

Since both np and np are greater than 5, use normal approximation with $\mu = np$ and $\sigma = \sqrt{npq}$.

$$P(180.5 \le x) = P\left(\frac{180.5 - 190.2}{8.7224} \le z\right) = P(-1.11 \le z) = 1 - P(z < -1.11) = 1 - 0.1335 = 0.8665$$

(b) $P(r < 200) = P(r \le 199) = P(x \le 199.5) = P\left(z \le \frac{199.5 - 190.2}{8.7224}\right) = P(z \le 1.07) = 0.8577$

(c) Let A be the event buy product; let B be the event tried free sample. Thus $P(\text{A, given B}) = 0.37$ and $P(\text{B}) = 0.60$. Since $P(\text{A and B}) = P(\text{B})$ ⊔ $P(\text{A, given B}) = 0.60(0.37) = 0.222$, $P(\text{sample and buy}) = 0.222$.

(d) Let a success be sample and buy. Then $p = 0.222$ (from (c)) and $q = 0.778$.

$$P(60 \le r \le 80) = P(59.5 \le x \le 80.5)$$

Here, $np = 317(0.222) = 70.374$ and $nq = 246.626$, so use normal approximation with $\mu = np$ and $\sigma = \sqrt{npq} = \sqrt{317(0.222)(0.778)} = \sqrt{54.750972} = 7.3994$.

$$\begin{aligned} P(59.5 \le x \le 80.5) &= P\left(\frac{59.5 - 70.374}{7.3994} \le z \le \frac{80.5 - 70.374}{7.3994}\right) \\ &= P(-1.47 \le z \le 1.37) \\ &= P(z \le 1.37) - P(z < -1.47) \\ &= 0.9147 - 0.0708 \\ &= 0.8439 \end{aligned}$$

15. $n = 267$ reservations, $P(\text{show}) = 1 - 0.06 = 0.94 = p$ so $q = 0.06$.

(a) $p = 0.94$

(b) Success = seat available for all who show up with a reservation, which means the number showing up must be ≤ 255 actual plane seats. Thus, $P(r \le 255)$.

(c) $P(r \le 255) = P(x \le 255.5)$

Since $np = 267(0.94) = 250.98 > 5$ and $nq = 267(0.06) = 16.02 > 5$, use normal approximation with $\mu = np$ and $\sigma = \sqrt{npq} = \sqrt{267(0.94)(0.06)} = \sqrt{15.0588} = 3.8806$.

$$P(x \le 255.5) = P\left(z \le \frac{255.5 - 250.98}{3.8806}\right) = P(z \le 1.16) = 0.8770$$

Chapter 6 Review

1. (a) $P(0 \le z \le 1.75) = P(z \le 1.75) - P(z < 0) = 0.9599 - 0.5 = 0.4599$

(b) $P(-1.29 \le z \le 0) = P(z \le 0) - P(z < -1.29) = 0.5 - 0.0985 = 0.4015$

(c) $P(1.03 \le z \le 1.21) = P(z \le 1.21) - P(z < 1.03) = 0.8869 - 0.8485 = 0.0384$

(d) $P(z \ge 2.31) = 1 - P(z < 2.31) = 1 - 0.9896 = 0.0104$

(e) $P(z \le -1.96) = 0.0250$

(f) $P(z \le 1) = 0.8413$

3. x is normal with $\mu = 47$ and $\sigma = 6.2$

(a) $P(x \le 60) = P\left(z \le \frac{60 - 47}{6.2}\right) = P(z \le 2.10) = 0.9821$

(b) $P(x \ge 50) = P\left(z \ge \frac{50 - 47}{6.2}\right) = P(z \ge 0.48) = 1 - P(z < 0.48) = 1 - 0.6844 = 0.3156$

(c) $P(50 \le x \le 60) = P(0.48 \le z \le 2.10) = P(z \le 2.10) - P(z < 0.48) = 0.9821 - 0.6844 = 0.2977$

5. Find z_0 such that $P(z \ge z_0) = 5\% = 0.05$. Same as find z_0 such that $P(z < z_0) = 0.95$

$P(z < 1.645) = 0.95$, so $z_0 = 1.645$

7. Find z_0 such that $P(-z_0 \le z \le +z_0) = 0.95$

Same as 5% of area outside $[-z_0, +z_0]$; split in half:

$P(z \le -z_0) = 0.05/2 = 0.025$

$P(z \le -1.96) = 0.0250$ so $-z_0 = -1.96$ and $+z_0 = 1.96$

9. $\mu = 79$, $\sigma = 9$

(a) $z = \frac{x - \mu}{\sigma} = \frac{87 - 79}{9} = 0.89$

(b) $z = \frac{79 - 79}{9} = 0$

(c) $P(x > 85) = P\left(z > \frac{85 - 79}{9}\right) = P(z > 0.67) = 1 - P(z \le 0.67) = 1 - 0.7486 = 0.2514$

11. Binomial with $n = 400$, $p = 0.70$, and $q = 0.30$.
Success = can recycled

(a) $P(r \ge 300) = P(300 \le r) = P(299.5 \le x)$

$np = 280 > 5,\ nq = 120 > 5,\ \sqrt{npq} = \sqrt{84} = 9.1652$

Use normal approximation with $\mu = np$ and $\sigma = \sqrt{npq}$.

$$P(299.5 \le x) = P\left(\frac{299.5-280}{9.1652} \le z\right) = P(2.13 \le z) = 1 - P(z < 2.13) = 1 - 0.9834 = 0.0166$$

(b)
$$\begin{aligned} P(260 \le r \le 300) &= P(259.5 \le x \le 300.5) \\ &= P\left(\frac{259.5-280}{9.1652} \le z \le \frac{300.5-280}{9.1652}\right) \\ &= P(-2.24 \le z \le 2.24) \\ &= P(z \le 2.24) - P(z < -2.24) \\ &= 0.9875 - 0.0125 \\ &= 0.9750 \end{aligned}$$

13. Delivery time x is normal with $\mu = 14$ and $\sigma = 2$ hours.

(a) $P(x \le 18) = P\left(z \le \frac{18-14}{2}\right) = P(z \le 2) = 0.9772$

(b) Find x_0 such that $P(x \le x_0) = 0.95$.
Find z_0 so that $P(z \le z_0) = 0.95$.
$P(z \le 1.645) = 0.95$, so $z_0 = 1.645$.
$x_0 = \mu + z_0\sigma = 14 + 1.645(2) = 17.29 \approx 17.3$ hours

15. Scanner price errors in the store's favor are mound-shaped with $\mu = \$2.66$ and $\sigma = \$0.85$.

(a) 68% of the errors should be in the range $\mu \pm 1\sigma$, approximately, or $2.66 \pm 1(0.85)$ which is \$1.81 to \$3.51.

(b) Approximately 95% of the errors should be in the range $\mu \pm 2\sigma$, or $2.66 \pm 2(0.85)$, which is \$0.96 to \$4.36.

(c) Almost all (99.7%) of the errors should lie in the range $\mu \pm 3\sigma$, or $2.66 \pm 3(0.85)$, which is \$0.11 to \$5.21.

17. Response time, x, is normally distributed with $\mu = 42$ and $\sigma = 8$ minutes.

(a)
$$\begin{aligned} P(30 \le x \le 45) &= P\left(\frac{30-42}{8} \le z \le \frac{45-42}{8}\right) \\ &= P(-1.5 \le z \le 0.375) \\ &= P(z \le 0.38) - P(z < -1.5) \\ &= 0.6480 - 0.0668 \\ &= 0.5812 \end{aligned}$$

(b) $P(x < 30) = P(z < -1.5) = 0.0668$

(c) $P(x > 60) = P\left(z > \frac{60-42}{8}\right) = P(z > 2.25) = 1 - P(z \le 2.25) = 1 - 0.9878 = 0.0122$

19. Success = having blood type AB

$n = 250, p = 3\% = 0.03, q = 1 - p = 0.97, np = 7.5, nq = 242.5, npq = 7.275$

(a) $P(5 \le r) = P(4.5 \le x)$

$np > 7.5$ and $\sigma = \sqrt{npq} = \sqrt{7.275} = 2.6972$

$$P(4.5 \le x) = P\left(\frac{4.5-7.5}{2.6972} \le z\right) = P(-1.11 \le z) = 1 - P(z < -1.11) = 1 - 0.1335 = 0.8665$$

(b)
$$\begin{aligned} P(5 \le r \le 10) &= P(4.5 \le x \le 10.5) \\ &= P\left(-1.11 \le z \le \frac{10.5-7.5}{2.6972}\right) \\ &= P(-1.11 \le z \le 1.11) \\ &= 1 - 2P(z < -1.11) \\ &= 1 - 2(0.1335) \\ &= 0.7330 \end{aligned}$$

Chapter 7 Introduction to Sampling Distributions

Section 7.1

1. Answers vary. Students should identify the individuals (subjects) and variable involved. Answers may include: A population is a set of measurements or counts either existing or conceptual. For example, the population of all ages of all people in Colorado; the population of weights of all students in your school; the population count of all antelope in Wyoming.

3. A population parameter is a numerical descriptive measure of a population, such as μ, the population mean; σ, the population standard deviation; σ^2, the population variance; p, the population proportion; and ρ (rho) the population correlation coefficient for those who have already studied linear regression from Chapter 10.

5. A statistical inference is a conclusion about the value of a population parameter based on information about the corresponding sample statistic and probability. We will do both estimation and testing.

7. They help us visualize the sampling distribution by using tables and graphs that approximately represent the sampling distribution.

9. We studied the sampling distribution of mean trout lengths based on samples of size 5. Other such sampling distributions abound. Notice that the sample size remains the same for each sample in a sampling distribution.

Section 7.2

Note: Answers may vary slightly depending on the number of digits carried in the standard deviation.

1. (a) $\mu_{\bar{x}} = \mu = 15$

$$\sigma_{\bar{x}} = \frac{\sigma}{\sqrt{n}} = \frac{14}{\sqrt{49}} = 2.0$$

Because $n = 49 \geq 30$, by the central limit theorem, we can assume that the distribution of $\bar{x}$ is approximately normal.

$$z = \frac{\bar{x} - \mu}{\sigma_{\bar{x}}} = \frac{\bar{x} - 15}{2.0}$$

$\bar{x} = 15$ converts to $z = \frac{15-15}{2.0} = 0$

$\bar{x} = 17$ converts to $z = \frac{17-15}{2.0} = 1$

$$\begin{aligned} P(15 \leq \bar{x} \leq 17) &= P(0 \leq z \leq 1) \\ &= P(z \leq 1) - P(z \leq 0) \\ &= 0.8413 - 0.5000 \\ &= 0.3413 \end{aligned}$$

(b) $\mu_{\bar{x}} = \mu = 15$

$$\sigma_{\bar{x}} = \frac{\sigma}{\sqrt{n}} = \frac{14}{\sqrt{64}} = 1.75$$

Because $n = 64 \geq 30$, by the central limit theorem, we can assume that the distribution of $\bar{x}$ is approximately normal.

$$z = \frac{\bar{x} - \mu}{\sigma_{\bar{x}}} = \frac{\bar{x} - 15}{1.75}$$

$\bar{x} = 15$ converts to $z = \dfrac{15 - 15}{1.75} = 0$

$\bar{x} = 17$ converts to $z = \dfrac{17 - 15}{1.75} = 1.14$

$$\begin{aligned} P(15 \leq \bar{x} \leq 17) &= P(0 \leq z \leq 1.14) \\ &= P(z \leq 1.14) - P(z \leq 0) \\ &= 0.8729 - 0.5000 \\ &= 0.3729 \end{aligned}$$

(c) The standard deviation of part (b) is smaller because of the larger sample size. Therefore, the distribution about $\mu_{\bar{x}}$ is narrower in part (b).

3. (a) No, we cannot say anything about the distribution of sample means because the sample size is only 9 and so it is too small to apply the central limit theorem.

(b) Yes, now we can say that the $\bar{x}$ distribution will also be normal with

$$\mu_{\bar{x}} = \mu = 25 \text{ and } \sigma_{\bar{x}} = \frac{\sigma}{\sqrt{n}} = \frac{3.5}{\sqrt{9}} = 1.17.$$

$$z = \frac{\bar{x} - \mu}{\sigma_{\bar{x}}} = \frac{\bar{x} - 25}{1.17}$$

$$\begin{aligned} P(23 \leq \bar{x} \leq 26) &= P\left(\frac{23 - 25}{1.17} \leq z \leq \frac{26 - 25}{1.17}\right) \\ &= P(-1.71 \leq z \leq 0.86) \\ &= P(z \leq 0.86) - P(z \leq -1.71) \\ &= 0.8051 - 0.0436 \\ &= 0.7615 \end{aligned}$$

5. (a) $\mu = 75,\ \sigma = 0.8$

$$\begin{aligned} P(x < 74.5) &= P\left(z < \frac{74.5 - 75}{0.8}\right) \\ &= P(z < -0.63) \\ &= 0.2643 \end{aligned}$$

(b) $\mu_{\bar{x}} = 75,\ \sigma_{\bar{x}} = \frac{\sigma}{\sqrt{n}} = \frac{0.8}{\sqrt{20}} = 0.179$

$$P(\bar{x} < 74.5) = P\left(z < \frac{74.5 - 75}{0.179}\right)$$
$$= P(z < -2.79)$$
$$= 0.0026$$

(c) No. If the weight of only one car were less than 74.5 tons, we cannot conclude that the loader is out of adjustment. If the mean weight for a sample of 20 cars were less than 74.5 tons, we would suspect that the loader is malfunctioning. As we see in part (b), the probability of this happening is very low if the loader is correctly adjusted.

7. (a) $\mu = 85,\ \sigma = 25$

$$P(x < 40) = P\left(z < \frac{40 - 85}{25}\right)$$
$$= P(z < -1.8)$$
$$= 0.0359$$

(b) The probability distribution of $\bar{x}$ is approximately normal with $\mu_{\bar{x}} = 85;\ \sigma_{\bar{x}} = \frac{\sigma}{\sqrt{n}} = \frac{25}{\sqrt{2}} = 17.68.$

$$P(\bar{x} < 40) = P\left(z < \frac{40 - 85}{17.68}\right)$$
$$= P(z < -2.55)$$
$$= 0.0054$$

(c) $\mu_{\bar{x}} = 85,\ \sigma_{\bar{x}} = \frac{\sigma}{\sqrt{n}} = \frac{25}{\sqrt{3}} = 14.43$

$$P(\bar{x} < 40) = P\left(z < \frac{40 - 85}{14.43}\right)$$
$$= P(z < -3.12)$$
$$= 0.0009$$

(d) $\mu_{\bar{x}} = 85,\ \sigma_{\bar{x}} = \frac{\sigma}{\sqrt{n}} = \frac{25}{\sqrt{5}} = 11.2$

$$P(\bar{x} < 40) = P\left(z < \frac{40 - 85}{11.2}\right)$$
$$= P(z < -4.02)$$
$$< 0.0002$$

(e) Yes; if the average value based on five tests were less than 40, the patient is almost certain to have excess insulin.

9. (a) $\mu = 63.0,\ \sigma = 7.1$

$$P(x < 54) = P\left(z < \frac{54 - 63.0}{7.1}\right)$$
$$= P(z < -1.27)$$
$$= 0.1020$$

(b) The expected number undernourished is 2200(0.1020) = 224.4, or about 224.

(c) $\mu_{\bar{x}} = 63.0,\ \sigma_{\bar{x}} = \dfrac{\sigma}{\sqrt{n}} = \dfrac{7.1}{\sqrt{50}} = 1.004$

$$\begin{aligned} P(\bar{x} < 60) &= P\left(z < \frac{60-63.0}{1.004}\right) \\ &= P(z < -2.99) \\ &= 0.0014 \end{aligned}$$

(d) $\mu_{\bar{x}} = 63.0,\ \sigma_{\bar{x}} = 1.004$

$$\begin{aligned} P(\bar{x} < 64.2) &= P\left(z < \frac{64.2-63.0}{1.004}\right) \\ &= P(z < 1.20) \\ &= 0.8849 \end{aligned}$$

Since the sample average is above the mean, it is quite unlikely that the doe population is undernourished.

11. (a) The random variable x is itself an average based on the number of stocks or bonds in the fund. Since x itself represents a sample mean return based on a large (random) sample of stocks or bonds, x has a distribution that is approximately normal (Central Limit Theorem).

(b) $\mu_{\bar{x}} = 1.6\%,\ \sigma_{\bar{x}} = \dfrac{\sigma}{\sqrt{n}} = \dfrac{0.9\%}{\sqrt{6}} = 0.367\%$

$$\begin{aligned} P(1\% \le \bar{x} \le 2\%) &= P\left(\frac{1\%-1.6\%}{0.367\%} \le z \le \frac{2\%-1.6\%}{0.367\%}\right) \\ &= P(-1.63 \le z \le 1.09) \\ &= P(z \le 1.09) - P(z \le -1.63) \\ &= 0.8621 - 0.0516 \\ &= 0.8105 \end{aligned}$$

Note: It does not matter whether you solve the problem using percents or their decimal equivalents as long as you are consistent.

(c) Note: 2 years = 24 months; x is <u>monthly</u> percentage return.

$\mu_{\bar{x}} = 1.6\%,\ \sigma_{\bar{x}} = \dfrac{\sigma}{\sqrt{n}} = \dfrac{0.9\%}{\sqrt{24}} = 0.1837\%$

$$\begin{aligned} P(1\% \le \bar{x} \le 2\%) &= P\left(\frac{1\%-1.6\%}{0.1837\%} \le z \le \frac{2\%-1.6\%}{0.1837\%}\right) \\ &= P(-3.27 \le z \le 2.18) \\ &= P(z \le 2.18) - P(z \le -3.27) \\ &= 0.9854 - 0.0005 \\ &= 0.9849 \end{aligned}$$

(d) Yes. The probability increases as the standard deviation decreases. The standard deviation decreases as the sample size increases.

(e) $\mu_{\bar{x}} = 1.6\%,\ \sigma_{\bar{x}} = 0.1837\%$

$$P(\bar{x} < 1\%) = P\left(z < \frac{1\% - 1.6\%}{0.1837\%}\right)$$
$$= P(z < -3.27)$$
$$= 0.0005$$

This is very unlikely if μ = 1.6%. One would suspect that μ has slipped below 1.6%.

13. (a) Since x itself represents a sample mean from a large $n \approx 80$ (random) sample of bonds, x is approximately normally distributed according to the Central Limit Theorem.

(b) $\mu_{\bar{x}} = 10.8\%,\ \sigma_{\bar{x}} = \frac{\sigma}{\sqrt{n}} = \frac{4.9\%}{\sqrt{5}} = 2.19\%$

$$P(\bar{x} < 6\%) = P\left(z < \frac{6\% - 10.8\%}{2.19\%}\right)$$
$$= P(z < -2.19)$$
$$= 0.0143$$

Yes. Since this probability is so small, it is very unlikely that $\bar{x}$ would be less than 6% if μ = 10.8%. The junk bond market appears to be weaker, i.e., μ is less than 10.8%.

(c) $\mu_{\bar{x}} = 10.8\%,\ \sigma_{\bar{x}} = 2.19\%$

$$P(\bar{x} > 16\%) = P\left(z > \frac{16\% - 10.8\%}{2.19\%}\right)$$
$$= P(z > 2.37)$$
$$= 1 - P(z \le 2.37)$$
$$= 1 - 0.9911$$
$$= 0.0089$$

Yes. Since this probability is so small, it is very unlikely that $\bar{x}$ would be greater than 16% if μ = 10.8%. The junk bond market may be heating up, i.e., μ is greater than 10.8%.

15. (a) The sample size should be 30 or more.

(b) No. If the distribution of x is normal, the distribution of $\bar{x}$ is also normal, regardless of the sample size.

17. (a) The total checkout time for 30 customers is the sum of the checkout times for each individual customer. Thus, $w = x_1 + x_2 + \cdots + x_{30}$ and the probability that the total checkout time for the next 30 customers is less than 90 is $P(w < 90)$.

(b) If we divide both sides of $w < 90$ by 30, we get $\frac{w}{30} < 3$. However, w is the sum of 30 waiting times, so $\frac{w}{30}$ is $\bar{x}$. Therefore, $P(w < 90) = P(\bar{x} < 3)$.

(c) The probability distribution of $\bar{x}$ is approximately normal with mean $\mu_{\bar{x}} = \mu = 2.7$ and standard deviation $\sigma_{\bar{x}} = \frac{\sigma}{\sqrt{n}} = \frac{0.6}{\sqrt{30}} = 0.1095$.

(d) $P(\bar{x} < 3) = P\left(z < \frac{3-2.7}{0.1095}\right)$

$= P(z < 2.74)$

$= 0.9969$

The probability that the total checkout time for the next 30 customers is less than 90 minutes is 0.9969, i.e., $P(w < 90) = 0.9969$.

19. Let $w = x_1 + x_2 + \ldots + x_{45}$.

(a) $w < 9500$ is equivalent to $\frac{w}{45} < \frac{9500}{45}$ or $\bar{x} < 211.111$. $\mu_{\bar{x}} = 240$, $\sigma_{\bar{x}} = \frac{\sigma}{\sqrt{n}} = \frac{84}{\sqrt{45}} = 12.522$

$$\begin{aligned} P(w < 9500) &= P(\bar{x} < 211.111) \\ &= P\left(z < \frac{211.111-240}{12.522}\right) \\ &= P(z < -2.31) \\ &= 0.0104 \end{aligned}$$

(b) $w < 12{,}000$ is equivalent to $\frac{w}{45} > \frac{12{,}000}{45}$ or $\bar{x} > 266.667$. $\mu_{\bar{x}} = 240$, $\sigma_{\bar{x}} = 12.522$

$$\begin{aligned} P(w > 12{,}000) &= P(\bar{x} > 266.667) \\ &= P\left(z > \frac{266.667-240}{12.522}\right) \\ &= P(z > 2.13) \\ &= 1 - P(z \le 2.13) \\ &= 1 - 0.9834 \\ &= 0.0166 \end{aligned}$$

(c)
$$\begin{aligned} P(9500 < w < 12{,}000) &= P(211.111 < \bar{x} < 266.667) \\ &= P(-2.31 < z < 2.13) \\ &= P(z < 2.13) - P(z < -2.31) \\ &= 0.9834 - 0.0104 \\ &= 0.9730 \end{aligned}$$

21. (a) Let $w = x_1 + x_2 + \cdots + x_5$.

$\mu_{\bar{x}} = \mu = 17$, $\sigma_{\bar{x}} = \frac{\sigma}{\sqrt{n}} = \frac{3.3}{\sqrt{5}} = 1.476$

$$\begin{aligned} P(w > 90) &= P\left(\frac{w}{5} > \frac{90}{5}\right) \\ &= P(\bar{x} > 18) \\ &= P\left(z > \frac{18-17}{1.476}\right) \\ &= P(z > 0.68) \\ &= 1 - 0.7517 \\ &= 0.2483 \end{aligned}$$

(b) $P(w < 80) = P\left(\frac{w}{5} < \frac{80}{5}\right)$

$$= P(\bar{x} < 16)$$

$$= P\left(z < \frac{16-17}{1.476}\right)$$

$$= P(z < -0.68)$$

$$= 0.2483$$

(c) $P(80 < w < 90) = P(16 < \bar{x} < 18)$

$$= P(-0.68 < z < 0.68)$$

$$= P(z < 0.68) - P(z < -0.68)$$

$$= 0.7517 - 0.2483$$

$$= 0.5034$$

Section 7.3

1. (a) Answers vary.

(b) The random variable $\hat{p}$ can be approximated by a normal random variable when both np and nq exceed 5.

$$\mu_{\hat{p}} = p, \sigma_{\hat{p}} = \sqrt{\frac{pq}{n}}$$

(c) $np = 33(0.21) = 6.93, nq = 33(0.79) = 26.07$

Yes, $\hat{p}$ can be approximated by a normal random variable since both np and nq exceed 5.

$$\mu_{\hat{p}} = p = 0.21, \sigma_{\hat{p}} = \sqrt{\frac{0.21(0.79)}{33}} \approx 0.071$$

$$\text{continuity correction} = \frac{0.5}{n} = \frac{0.5}{33} \approx 0.015$$

$$P(0.15 \le \hat{p} \le 0.25) = P(0.15 - 0.015 \le x \le 0.25 + 0.015)$$

$$= P(0.135 \le x \le 0.265)$$

$$= P\left(\frac{0.135-0.21}{0.071} \le z \le \frac{0.265-0.21}{0.071}\right)$$

$$= P(-1.06 \le z \le 0.77)$$

$$= P(z \le 0.77) - P(z \le -1.06)$$

$$= 0.7794 - 0.1446$$

$$= 0.6348$$

(d) No; $np = 25(0.15) = 3.75$, which does not exceed 5.

(e) $np = 48(0.15) = 7.2$, $nq = 48(0.85) = 40.8$

Yes, $\hat{p}$ can be approximated by a normal random variable since both np and nq exceed 5.

$$\mu_{\hat{p}} = p = 0.15,\ \sigma_{\hat{p}} = \sqrt{\frac{0.15(0.85)}{48}} \approx 0.052$$

$$\text{continuity correction} = \frac{0.5}{n} = \frac{0.5}{45} = 0.010$$

$$\begin{aligned} P(\hat{p} \ge 0.22) &= P(x \ge 0.22 - 0.010) \\ &= P(x \ge 0.21) \\ &= P\left(z \ge \frac{0.21 - 0.15}{0.052}\right) \\ &= P(z \ge 1.15) \\ &= 1 - P(z < 1.15) \\ &= 1 - 0.8749 \\ &= 0.1251 \end{aligned}$$

3. $n = 30$, $p = 0.60$

$np = 30(0.60) = 18$, $nq = 30(0.40) = 12$

Approximate $\hat{p}$ by a normal random variable since both np and nq exceed 5.

$$\mu_{\hat{p}} = p = 0.6,\ \sigma_{\hat{p}} = \sqrt{\frac{0.6(0.4)}{30}} \approx 0.089$$

$$\text{continuity correction} = \frac{0.5}{n} = \frac{0.5}{30} = 0.017$$

(a)
$$\begin{aligned} P(\hat{p} \ge 0.5) &\approx P(x \ge 0.5 - 0.017) \\ &= P(x \ge 0.483) \\ &= P\left(z \ge \frac{0.483 - 0.6}{0.089}\right) \\ &= P(z \ge -1.31) \\ &= 0.9049 \end{aligned}$$

(b)
$$\begin{aligned} P(\hat{p} \ge 0.667) &\approx P(x \ge 0.667 - 0.017) \\ &= P(x \ge 0.65) \\ &= P\left(z \ge \frac{0.65 - 0.6}{0.089}\right) \\ &= P(z \ge 0.56) \\ &= 0.2877 \end{aligned}$$

(c) $P(\hat{p} \le 0.333) \approx P(x \le 0.333 + 0.017)$

$$= P(x \le 0.35)$$
$$= P\left(z \le \frac{0.35 - 0.6}{0.089}\right)$$
$$= P(z \le -2.81)$$
$$= 0.0025$$

(d) Yes, both np and nq exceed 5.

5. $n = 55,\ p = 0.11$

$np = 55(0.11) = 6.05,\ nq = 55(0.89) = 48.95$

Approximate $\hat{p}$ by a normal random variable since both np and nq exceed 5.

$$\mu_{\hat{p}} = p = 0.11,\ \sigma_{\hat{p}} = \sqrt{\frac{0.11(0.89)}{55}} \approx 0.042$$

$$\text{continuity correction} = \frac{0.5}{n} = \frac{0.5}{55} = 0.009$$

(a) $P(\hat{p} \le 0.15) \approx P(x \le 0.15 + 0.009)$

$$= P(x \le 0.159)$$
$$= P\left(z \le \frac{0.159 - 0.11}{0.042}\right)$$
$$= P(z \le 1.17)$$
$$= 0.8790$$

(b) $P(0.10 \le \hat{p} \le 0.15) \approx P(0.10 - 0.009 \le x \le 0.15 + 0.009)$

$$= P(0.091 \le x \le 0.159)$$
$$= P\left(\frac{0.091 - 0.11}{0.042} \le z \le \frac{0.159 - 0.11}{0.042}\right)$$
$$= P(-0.45 \le z \le 1.17)$$
$$= P(z \le 1.17) - P(z \le -0.45)$$
$$= 0.8790 - 0.3264$$
$$= 0.5526$$

(c) Yes, both np and nq exceed 5.

7. (a) $n = 100,\ p = 0.06$

$np = 100(0.06) = 6,\ nq = 100(0.94) = 94$

$\hat{p}$ can be approximated by a normal random variable since both np and nq exceed 5.

$$\mu_{\hat{p}} = p = 0.06,\ \sigma_{\hat{p}} = \sqrt{\frac{0.06(0.94)}{100}} \approx 0.024$$

$$\text{continuity correction} = \frac{0.5}{100} = 0.005$$

(b) $P(\hat{p} \geq 0.07) \approx P(x \geq 0.07 - 0.005)$

$= P(x \geq 0.065)$

$= P\left(z \geq \dfrac{0.065 - 0.06}{0.024}\right)$

$= P(z \geq 0.21)$

$= 0.4168$

(c) $P(\hat{p} \geq 0.11) \approx P(x \geq 0.11 - 0.005)$

$= P(x \geq 0.105)$

$= P\left(z \geq \dfrac{0.105 - 0.06}{0.024}\right)$

$= P(z \geq 1.88)$

$= 0.0301$

Yes, since this probability is so small, it should rarely occur. The machine might need an adjustment.

9. $\bar{p} = \dfrac{\text{total number of successes from all 12 quarters}}{\text{total number of families from all 12 quarters}}$

$= \dfrac{11 + 14 + \ldots + 19}{12(92)}$

$= \dfrac{206}{1104}$

$= 0.1866$

$\bar{q} = 1 - \bar{p} = 1 - 0.1866 = 0.8134$

$\mu_{\hat{p}} = p \approx \bar{p} = 0.1866$

$\sigma_{\hat{p}} = \sqrt{\dfrac{pq}{n}} \approx \sqrt{\dfrac{\bar{p}\bar{q}}{n}} = \sqrt{\dfrac{0.1866(0.8134)}{92}} \approx 0.0406$

Check: $n\bar{p} = 92(0.1866) = 17.2$, $n\bar{q} = 92(0.8134) = 74.8$

Since both $n\bar{p}$ and $n\bar{q}$ exceed 5, the normal approximation should be reasonably good.

Center line = $\bar{p} = 0.1866$

Control limits at $\bar{p} \pm 2\sqrt{\dfrac{\bar{p}\bar{q}}{n}}$

$= 0.1866 \pm 2(0.0406)$

$= 0.1866 \pm 0.0812$

or 0.1054 and 0.2678

Control limits at $\bar{p} \pm 3\sqrt{\dfrac{\bar{p}\bar{q}}{n}}$

$= 0.1866 \pm 3(0.0406)$

$= 0.1866 \pm 0.1218$

or 0.0648 and 0.3084

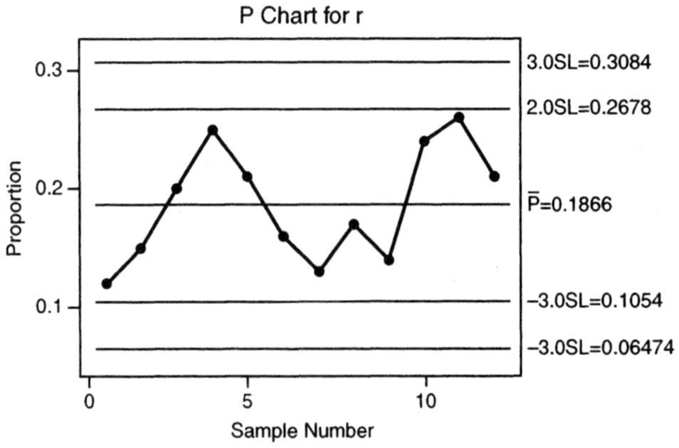

There are no out-of-control signals.

11. $\bar{p} = \dfrac{\text{total number who got jobs}}{\text{total number of people}}$

$= \dfrac{60+53+\ldots+58}{75(15)}$

$= \dfrac{872}{1125}$

$= 0.7751$

$\bar{q} = 1-\bar{p} = 1-0.7751 = 0.2249$

$\mu_{\hat{p}} = p \approx \bar{p} = 0.7751$

$\sigma_{\hat{p}} = \sqrt{\dfrac{pq}{n}} \approx \sqrt{\dfrac{\bar{p}\bar{q}}{n}} = \sqrt{\dfrac{(0.7751)(0.2249)}{75}} \approx 0.0482$

Check: $n\bar{p} = 75(0.7751) = 58.1$, $n\bar{q} = 75(0.2249) = 16.9$

Since both $n\bar{p}$ and $n\bar{q}$ exceed 5, the normal approximation should be reasonably good.

Center line $= \bar{p} = 0.7751$

Control limits at $\bar{p} \pm 2\sqrt{\dfrac{\bar{p}\bar{q}}{n}}$

$= 0.7751 \pm 2(0.0482)$

$= 0.7751 \pm 0.0964$

or 0.6787 to 0.8715

Control limits at $\bar{p} \pm 3\sqrt{\dfrac{\bar{p}\bar{q}}{n}}$

$= 0.7751 \pm 3(0.0482)$

$= 0.7751 \pm 0.1446$

or 0.6305 to 0.9197

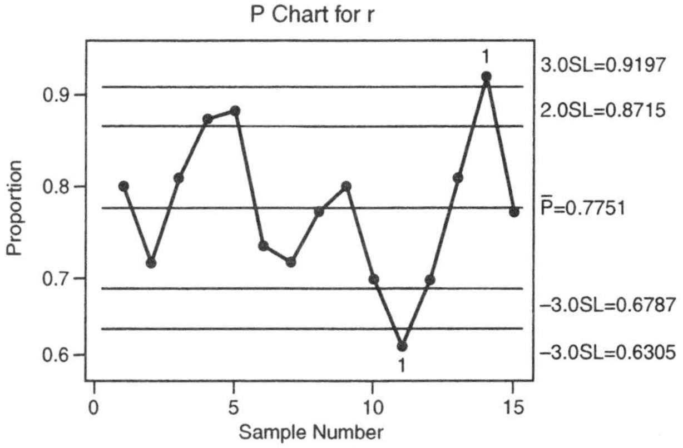

Out-of-control signal III occurs on days 4 and 5, out-of-control signal I occurs on day 11 on the low side and day 14 on the high side. Out-of-control signals on the low side are of most concern for the homeless seeking work. The foundation should look to see what happened on that day. The foundation might take a look at the out-of-control periods on the high side to see if there is a possibility of cultivating more jobs.

Chapter 7 Review

1. (a) The $\bar{x}$ distribution approaches a normal distribution.

(b) The mean $\mu_{\bar{x}}$ of the $\bar{x}$ distribution equals the mean μ of the x distribution, regardless of the sample size.

(c) The standard deviation $\sigma_{\bar{x}}$ of the sampling distribution equals $\dfrac{\sigma}{\sqrt{n}}$, where σ is the standard deviation of the x distribution and n is the sample size.

(d) They will both be approximately normal with the same mean, but the standard deviations will be $\dfrac{\sigma}{\sqrt{50}}$ and $\dfrac{\sigma}{\sqrt{100}}$ respectively.

3. (a) $\mu = 35,\ \sigma = 7$

$$P(x \ge 40) = P\left(z \ge \frac{40-35}{7}\right)$$
$$= P(z \ge 0.71)$$
$$= 0.2389$$

(b) $\mu_{\bar{x}} = \mu = 35,\ \sigma_{\bar{x}} = \dfrac{\sigma}{\sqrt{n}} = \dfrac{7}{\sqrt{9}} = \dfrac{7}{3}$

$$P(\bar{x} \ge 40) = P\left(z \ge \frac{40-35}{\frac{7}{3}}\right)$$
$$= P(z \ge 2.14)$$
$$= 0.0162$$

5. $\mu_{\bar{x}} = \mu = 100,\ \sigma_{\bar{x}} = \dfrac{\sigma}{\sqrt{n}} = \dfrac{15}{\sqrt{100}} = 1.5$

$$\begin{aligned} P(100-2 \le \bar{x} \le 100+2) &= P(98 \le \bar{x} \le 102) \\ &= P\left(\frac{98-100}{1.5} \le z \le \frac{102-100}{1.5}\right) \\ &= P(-1.33 \le z \le 1.33) \\ &= P(z \le 1.33) - P(z \le -1.33) \\ &= 0.9082 - 0.0918 \\ &= 0.8164 \end{aligned}$$

7. $\mu_{\bar{x}} = \mu = 750,\ \sigma_{\bar{x}} = \dfrac{\sigma}{\sqrt{n}} = \dfrac{20}{\sqrt{64}} = 2.5$

(a) $$\begin{aligned} P(\bar{x} \ge 750) &= P\left(z \ge \frac{750-750}{2.5}\right) \\ &= P(z \ge 0) \\ &= 0.5000 \end{aligned}$$

(b) $$\begin{aligned} P(745 \le \bar{x} \le 755) &= P\left(\frac{745-750}{2.5} \le z \le \frac{755-750}{2.5}\right) \\ &= P(-2 \le z \le 2) \\ &= P(z \le 2) - P(z \le -2) \\ &= 0.9772 - 0.0228 \\ &= 0.9544 \end{aligned}$$

9. (a) $n = 50,\ p = 0.22$

$np = 50(0.22) = 11,\ nq = 50(0.78) = 39$

Approximate $\hat{p}$ by a normal random variable since both np and nq exceed 5.

$$\mu_{\hat{p}} = p = 0.22,\ \sigma_{\hat{p}} = \sqrt{\frac{0.22(0.78)}{50}} \approx 0.0586$$

$$\text{continuity correction} = \frac{0.5}{n} = \frac{0.5}{50} = 0.01$$

$$\begin{aligned} P(0.20 \le \hat{p} \le 0.25) &\approx P(0.20 - 0.01 \le x \le 0.25 + 0.01) \\ &= P(0.19 \le x \le 0.26) \\ &= P\left(\frac{0.19-0.22}{0.0586} \le z \le \frac{0.26-0.22}{0.0586}\right) \\ &= P(-0.51 \le z \le 0.68) \\ &= P(z \le 0.68) - P(z \le -0.51) \\ &= 0.7517 - 0.3050 \\ &= 0.4467 \end{aligned}$$

(b) $n = 38,\ p = 0.27$

$np = 38(0.27) = 10.26,\ nq = 38(0.73) = 27.74$

Approximate $\hat{p}$ by a normal random variable since both np and nq exceed 5.

$$\mu_{\hat{p}} = p = 0.27,\ \sigma_{\hat{p}} = \sqrt{\frac{(0.27)(0.73)}{38}} \approx 0.0720$$

$$\text{continuity correction} = \frac{0.5}{n} = \frac{0.5}{38} = 0.013$$

$$\begin{aligned} P(\hat{p} \geq 0.35) &\approx P(x \geq 0.35 - 0.013) \\ &= P(x \geq 0.337) \\ &= P\left(z \geq \frac{0.337 - 0.27}{0.0720}\right) \\ &= P(z \geq 0.93) \\ &= 0.1762 \end{aligned}$$

(c) $n = 51,\ p = 0.05$

$np = 51(0.05) = 2.55$

No, we cannot approximate $\hat{p}$ by a normal random variable since $np < 5$.

Chapter 8 Estimation

Section 8.1

Answers may vary slightly due to rounding.

1. (a) $n = 15,\ \overline{x} = 3.15,\ \sigma = 0.33,\ c = 80\%,\ z_c = 1.28$

$$E = \frac{z_c\sigma}{\sqrt{n}} = \frac{1.28(0.33)}{\sqrt{15}} \approx 0.11$$

$$(\overline{x} - E) < \mu < (\overline{x} + E)$$
$$3.15 - 0.11 < \mu < 3.15 + 0.11$$
$$3.04 \text{ g} < \mu < 3.26 \text{ g}$$

The margin of error, E, is 0.11 g.

(b) The distribution of weights is assumed to be normal and σ is known.

(c) There is an 80% chance the confidence interval is one of the intervals that contains the population average weight of Allen's hummingbirds in this region.

3. (a) $n = 45,\ \overline{x} = 37.5,\ \sigma = 7.50,\ c = 99\%,\ z_c = 2.58$

$$E = \frac{z_c\sigma}{\sqrt{n}} = \frac{2.58(7.50)}{\sqrt{45}} \approx 2.88$$

$$(\overline{x} - E) < \mu < (\overline{x} + E)$$
$$37.5 - 2.88 < \mu < 37.5 + 2.88$$
$$34.62 \text{ ml/kg} < \mu < 40.38 \text{ ml/kg}$$

The margin of error, E, is 2.88 ml/kg.

(b) The sample size is large (30 or more) and σ is known.

(c) There is a 99% chance the confidence interval is one of the intervals that contains the population average blood plasma level for male firefighters.

5. $n = 30,\ \overline{x} = 138.5,\ \sigma = 42.6$

(a) $c = 90\%,\ z_c = 1.645$

$$E = \frac{z_c\sigma}{\sqrt{n}} = \frac{1.645(42.6)}{\sqrt{30}} \approx 12.8$$
$$(\overline{x} - E) < \mu < (\overline{x} + E)$$
$$(138.5 - 12.8) < \mu < (138.5 + 12.8)$$
$$125.7 \text{ larceny cases} < \mu < 151.3 \text{ larceny cases}$$

The margin of error, E, is 12.8 larceny cases.

(b) $c = 95\%,\ z_c = 1.96$

$$E = \frac{z_c\sigma}{\sqrt{n}} = \frac{1.96(42.6)}{\sqrt{30}} \approx 15.2$$
$$(\overline{x} - E) < \mu < (\overline{x} + E)$$
$$(138.5 - 15.2) < \mu < (138.5 + 15.2)$$
$$123.3 \text{ larceny cases} < \mu < 153.7 \text{ larceny cases}$$

The margin of error, E, is 15.2 larceny cases.

(c) $c = 99\%,\ z_c = 2.58$

$$E = \frac{z_c\sigma}{\sqrt{n}} = \frac{2.58(42.6)}{\sqrt{30}} \approx 20.1$$

$$(\overline{x} - E) < \mu < (\overline{x} + E)$$

$$(138.5 - 20.1) < \mu < (138.5 + 20.1)$$

118.4 larceny cases $< \mu < 158.6$ larceny cases

The margin of error, E, is 20.1 larceny cases.

(d) Yes.

(e) Yes.

7. $\overline{x} = 58{,}940,\ \sigma = 18{,}490,\ c = 90\%,\ z_c = 1.645$

(a) $n = 36$

$$E = \frac{z_c\sigma}{\sqrt{n}} = \frac{1.645(18{,}490)}{\sqrt{36}} \approx 5069$$

$$(\overline{x} - E) < \mu < (\overline{x} + E)$$

$$58{,}940 - 5069 < \mu < 58{,}940 + 5069$$

$$\$53{,}871 < \mu < \$64{,}009$$

The margin of error, E, is \$5069.

(b) $n = 64$

$$E = \frac{z_c\sigma}{\sqrt{n}} = \frac{1.645(18{,}490)}{\sqrt{64}} \approx 3802$$

$$(\overline{x} - E) < \mu < (\overline{x} + E)$$

$$(58{,}940 - 3802) < \mu < (58{,}940 + 3802)$$

$$\$55{,}138 < \mu < \$62{,}742$$

The margin of error, E, is \$3802.

(c) $n = 121$

$$E = \frac{z_c\sigma}{\sqrt{n}} = \frac{1.645(18{,}490)}{\sqrt{121}} \approx 2765$$

$$(\overline{x} - E) < \mu < (\overline{x} + E)$$

$$(58{,}940 - 2765) < \mu < (58{,}940 + 2765)$$

$$\$56{,}175 < \mu < \$61{,}705$$

The margin of error, E, is \$2765.

(d) Yes.

(e) Yes.

9. (a) $n = 42,\ \overline{x} = \frac{\sum x_i}{n} = \frac{1511.8}{42} = 35.9952 \approx 36.0,$ as stated.

(b) $c = 75\%,\ z_c = 1.15$

$$E \approx \frac{z_c\sigma}{\sqrt{n}} = \frac{1.15(10.2)}{\sqrt{42}} = 1.81$$

$$(\overline{x} - E) < \mu < (\overline{x} + E)$$

$$(36.0 - 1.81) < \mu < (36.0 + 1.81)$$

$34.19 < \mu < 37.81$ thousand dollars per employee profit

(c) Yes. Since \$30 thousand per employee profit is less than the lower limit of the confidence interval (34.19), your bank profits are low compared to other similar financial institutions.

(d) Yes. Since $40 thousand per employee profit exceeds the upper limit of the confidence interval (37.81), your bank profit is higher than other similar financial institutions.

(e) $c = 90\%,\ z_c = 1.645$

$$E \approx \frac{z_c \sigma}{\sqrt{n}} = \frac{1.645(10.2)}{\sqrt{42}} = 2.59$$

$$(\bar{x} - E) < \mu < (\bar{x} + E)$$
$$(36.0 - 2.59) < \mu < (36.0 + 2.59)$$

$33.41 < \mu < 38.59$ thousand dollars per employee profit

Yes. $30 thousand is less than the lower limit of the confidence interval (33.41), so your bank's profit is less than that of other financial institutions.
Yes. $40 thousand is more than the upper limit of the confidence interval (38.59), so your bank is doing better (profit-wise) than other financial institutions.

11. (a) $n = 40,\ \bar{x} = 51.1575 \approx 51.16$, as stated.

(b) $c = 90\%,\ z_c = 1.645$

$$E \approx \frac{z_c \sigma}{\sqrt{n}} = \frac{1.645(3.04)}{\sqrt{40}} = 0.79$$

$$(\bar{x} - E) < \mu < (\bar{x} + E)$$
$$(51.16 - 0.79) < \mu < (51.16 + 0.79)$$
$$50.37°\text{F} < \mu < 51.95°\text{F}$$

(c) $c = 99\%,\ z_c = 2.58$

$$E \approx \frac{z_c \sigma}{\sqrt{n}} = \frac{2.58(3.04)}{\sqrt{40}} = 1.24$$

$$(\bar{x} - E) < \mu < (\bar{x} + E)$$
$$(51.16 - 1.24) < \mu < (51.16 + 1.24)$$
$$49.92°\text{F} < \mu < 52.40°\text{F}$$

(d) Since 53°F > 52.4°F, the upper confidence interval limit, it is unlikely the average January temperature is 53°F. Plotting the data by year to see if there is an upward trend (lately or overall), and/or adding several more years of observation might provide evidence to support or refute the claim.

Section 8.2

Answers may vary slightly due to rounding.

1. $n = 18$ so $d.f. = n - 1 = 18 - 1 = 17$, $c = 0.95$
$t_c = t_{0.95} = 2.110$

3. $n = 22$ so $d.f. = n - 1 = 22 - 1 = 21$, $c = 0.90$
$t_c = t_{0.90} = 1.721$

5. $n = 9$ so $d.f. = n - 1 = 9 - 1 = 8$

(a) $\bar{x} = \dfrac{\sum x}{n} = \dfrac{11,450}{9} \approx 1272$, as stated

$$s^2 = \frac{\sum x_i^2 - \frac{(\sum x_i)^2}{n}}{n-1} = \frac{14,577,854 - \frac{(11,450)^2}{9}}{8} = 1363.6944$$

$s = \sqrt{1363.6944} = 36.9282 \approx 37$, as stated

(b) $c = 90\%,\ t_c = t_{0.90}$ with $8\ d.f. = 1.860$

$$E = \frac{t_c s}{\sqrt{n}} = \frac{1.86(37)}{\sqrt{9}} = 22.94 \approx 23$$

$(\overline{x} - E) < \mu < (\overline{x} + E)$

$(1272 - 23) < \mu < (1272 + 23)$

1249 A.D. $< \mu <$ 1295 A.D.

7. $n = 6$ so $d.f. = n - 1 = 5$

(a) $\overline{x} = 91.0$, as stated

$s = 30.7181 \approx 30.7$, as stated

(b) $c = 75\%$, so $t_{0.75}$ with $5\ d.f. = 1.301$

$$E = \frac{t_c s}{\sqrt{n}} = \frac{1.301(30.7)}{\sqrt{6}} \approx 16.3$$

$(\overline{x} - E) < \mu < (\overline{x} + E)$

$(91.0 - 16.3) < \mu < (91.0 + 16.3)$

74.7 pounds $< \mu <$ 107.3 pounds

9. $n = 6$ so $d.f. = n - 1 = 5$

$\overline{x} = 79.25,\ s = 5.33$

$c = 80\%$, so $t_{0.80}$ with $5\ d.f. = 1.476$

$$E = \frac{t_c s}{\sqrt{n}} = \frac{1.476(5.33)}{\sqrt{6}} \approx 3.21$$

$(\overline{x} - E) < \mu < (\overline{x} + E)$

$(79.25 - 3.21) < \mu < (79.25 + 3.21)$

76.04 cm $< \mu <$ 82.46 cm

11. $n = 10$, so $d.f. = n - 1 = 9$

(a) $\overline{x} = 9.9500 \approx 9.95$, as stated.

$s \approx 1.0212 \approx 1.02$, as stated.

(b) $c = 99.9\%$, so $t_{0.999}$ with $9\ d.f. = 4.781$

$$E = \frac{t_c s}{\sqrt{n}} = \frac{4.781(1.02)}{\sqrt{10}} \approx 1.54$$

$(\overline{x} - E) < \mu < (x + E)$

$9.95 - 1.54 < \mu < 9.95 + 1.54$

8.41 mg/dl $< \mu <$ 11.49 mg/dl

(c) Since all values in the 99.9% confidence interval are above 6 mg/dl, we can be almost certain that this patient does not have a calcium deficiency.

13. **(a)** $n = 19$, $d.f. = n - 1 = 18$, $c = 90\%$, $t_{0.90}$ with 18 $d.f. = 1.734$

$\bar{x} = 9.8421 \approx 9.8$, as stated

$s = 3.3001 \approx 3.3$, as stated

$$E = \frac{t_c s}{\sqrt{n}} = \frac{1.734(3.3)}{\sqrt{19}} \approx 1.3$$

$(\bar{x} - E) < \mu < (\bar{x} + E)$

$(9.8 - 1.3) < \mu < (9.8 + 1.3)$

8.5 inches $< \mu <$ 11.1 inches

(b) $n = 7$, $d.f. = n - 1 = 6$, $c = 80\%$, $t_{0.80}$ with 6 $d.f. = 1.440$

$\bar{x} = 17.0714 \approx 17.1$, as stated

$s = 2.4568 \approx 2.5$, as stated

$$E = \frac{t_c s}{\sqrt{n}} = \frac{1.440(2.5)}{\sqrt{7}} \approx 1.4$$

$(\bar{x} - E) < \mu < (\bar{x} + E)$

$(17.1 - 1.4) < \mu < (17.1 + 1.4)$

15.7 inches $< \mu <$ 18.5 inches

15. Notice that the four figures are drawn on different scales.

(a) The box plots differ in range (distance between whisker ends), in interquartile range (distance between box ends), in medians (line through boxes), in symmetry (indicated by the placement of the median within the box, and by the placement of the median relative to the whisker endpoints), in whisker lengths, and in the presence/absence of outliers. These differences are to be expected since each box plot represents a different sample of size 20. (Although the data sets were all selected as samples of size $n = 20$ from a normal distribution with $\mu = 68$ and $\sigma = 3$, it is very interesting that Sample 2, figure (b), shows 2 outliers.)

(b)

Sample	Confidence interval width	Includes $\mu = 68$?
1	69.407 – 66.692 = 2.715	Yes
2	69.426 – 66.490 = 2.936	Yes
3	69.211 – 66.741 = 2.470	Yes
4	68.050 – 65.766 = 2.284	Yes (barely)

The intervals differ in length; all four enclose $\mu = 68$. If many additional samples of size 20 were generated from this distribution, we would expect about 95% of the confidence intervals created from these samples to enclose the number 68; in approximately 5% of the intervals, 68 would be outside the confidence interval, i.e., 68 would be less than the lower limit or greater than the upper limit. Drawing all these samples (at least conceptually), checking whether or not the confidence intervals include the number 68, and keeping track of the percentage that do, is an illustration of the definition of (95%) confidence intervals.

17. **(a)** $\overline{x} = 25.176 \approx 25.2$, as stated
$s = 15.472 \approx 15.5$, as stated

(b) $n = 51$, so $d.f. = n - 1 = 50$
$c = 90\%$, $t_{0.90}$ with 50 $d.f. = 1.676$

$$E = \frac{t_c s}{\sqrt{n}} = \frac{1.676(15.5)}{\sqrt{51}} = 3.6$$
$$(\overline{x} - E) < \mu < (\overline{x} + E)$$
$$(25.2 - 3.6) < \mu < (25.2 + 3.6)$$
$$21.6 < \mu < 28.8$$

(c) $c = 99\%$, $t_{0.99}$ with 50 $d.f. = 2.678$

$$E = \frac{t_c s}{\sqrt{n}} = \frac{2.678(15.5)}{\sqrt{51}} \approx 5.8$$
$$(\overline{x} - E) < \mu < (\overline{x} + E)$$
$$(25.2 - 5.8) < \mu < (25.2 + 5.8)$$
$$19.4 < \mu < 31.0$$

(d) Using both confidence intervals, we can say that the P/E for Bank One is well below the population average. The P/E for AT&T Wireless is well above the population average. The P/E for Disney is within both confidence intervals. It appears that the P/E for Disney is close to the population average P/E.

(e) By the central limit theorem, when n is large the $\overline{x}$ distribution is approximately normal. In general, $n \geq 30$ is considered large.

19. **(a)** $n = 31$, $d.f. = n - 1 = 30$, $\overline{x} = 45.2$, $s = 5.3$

$c = 90\%$, $t_{0.90} = 1.697$, $E = \dfrac{t_c s}{\sqrt{n}} = \dfrac{1.697(5.3)}{\sqrt{31}} \approx 1.62$

$c = 95\%$, $t_{0.95} = 2.042$, $E = \dfrac{t_c s}{\sqrt{n}} = \dfrac{2.042(5.3)}{\sqrt{31}} \approx 1.94$

$c = 99\%$, $t_{0.99} = 2.750$, $E = \dfrac{t_c s}{\sqrt{n}} = \dfrac{2.750(5.3)}{\sqrt{31}} \approx 2.62$

90% C.I.: $(45.2 - 1.62) < \mu < (45.2 + 1.62)$
$43.58 < \mu < 46.82$

95% C.I.: $(45.2 - 1.94) < \mu < (45.2 + 1.94)$
$43.26 < \mu < 47.14$

99% C.I.: $(45.2 - 2.62) < \mu < (45.2 + 2.62)$
$42.58 < \mu < 47.82$

(b) $n = 31,\ \bar{x} = 45.2,\ \sigma = 5.3$

$c = 90\%,\ z_{0.90} = 1.645,\ E = \dfrac{z_c\sigma}{\sqrt{n}} = \dfrac{1.645(5.3)}{\sqrt{31}} \approx 1.57$

$c = 95\%,\ z_{0.95} = 1.96,\ E = \dfrac{z_c\sigma}{\sqrt{n}} = \dfrac{1.96(5.3)}{\sqrt{31}} \approx 1.87$

$c = 99\%,\ z_{0.99} = 2.58,\ E = \dfrac{z_c\sigma}{\sqrt{n}} = \dfrac{2.58(5.3)}{\sqrt{31}} \approx 2.46$

90% C.I.: $(45.2 - 1.57) < \mu < (45.2 + 1.57)$

$43.63 < \mu < 46.77$

95% C.I.: $(45.2 - 1.87) < \mu < (45.2 + 1.87)$

$43.33 < \mu < 47.07$

99% C.I.: $(45.2 - 2.46) < \mu < (45.2 + 2.46)$

$42.74 < \mu < 47.66$

(c) Yes; the respective intervals based on the Student's t distribution are slightly longer.

(d) For Student's t, $n = 81$, $d.f. = n - 1 = 80$, $\bar{x} = 45.2,\ s = 5.3$

$c = 90\%,\ t_{0.90} = 1.664,\ E = \dfrac{t_c s}{\sqrt{n}} = \dfrac{1.664(5.3)}{\sqrt{81}} \approx 0.98$

$c = 95\%,\ t_{0.95} = 1.990,\ E = \dfrac{t_c s}{\sqrt{n}} = \dfrac{1.990(5.3)}{\sqrt{81}} \approx 1.17$

$c = 99\%,\ t_{0.99} = 2.639,\ E = \dfrac{t_c s}{\sqrt{n}} = \dfrac{2.639(5.3)}{\sqrt{81}} \approx 1.55$

90% C.I.: $(45.2 - 0.98) < \mu < (45.2 + 0.98)$

$44.22 < \mu < 46.18$

95% C.I.: $(45.2 - 1.17) < \mu < (45.2 + 1.17)$

$44.03 < \mu < 46.37$

99% C.I.: $(45.2 - 1.55) < \mu < (45.2 + 1.55)$

$43.65 < \mu < 46.75$

For standard normal, $n = 81,\ \bar{x} = 45.2,\ \sigma = 5.3$

$c = 90\%,\ z_{0.90} = 1.645,\ E = \dfrac{z_c\sigma}{\sqrt{n}} = \dfrac{1.645(5.3)}{\sqrt{81}} \approx 0.97$

$c = 95\%,\ z_{0.95} = 1.96,\ E = \dfrac{z_c\sigma}{\sqrt{n}} = \dfrac{1.96(5.3)}{\sqrt{81}} \approx 1.15$

$c = 99\%,\ z_{0.99} = 2.58,\ E = \dfrac{z_c\sigma}{\sqrt{n}} = \dfrac{2.58(5.3)}{\sqrt{81}} \approx 1.52$

90% C.I.: $(45.2 - 0.97) < \mu < (45.2 + 0.97)$

$44.23 < \mu < 46.17$

95% C.I.: $(45.2 - 1.15) < \mu < (45.2 + 1.15)$

$44.05 < \mu < 46.35$

99% C.I.: $(45.2 - 1.52) < \mu < (45.2 + 1.52)$

$43.68 < \mu < 46.72$

The intervals using the t distribution are still slightly longer than corresponding intervals using the standard normal distribution. However, with a larger sample size, the differences between the two methods is less pronounced.

Section 8.3

Answers may vary slightly due to rounding.

1. $r = 39,\ n = 62,\ \hat{p} = \frac{r}{n} = \frac{39}{62},\ \hat{q} = 1 - \hat{p} = \frac{23}{62}$

(a) $\hat{p} = \frac{39}{62} = 0.6290$

(b) $c = 95\%, z_c = z_{0.95} = 1.96$
$E \approx z_c\sqrt{\hat{p}\hat{q}/n} = 1.96\sqrt{(0.6290)(1-0.6290)/62} = 0.1202$
$(\hat{p} - E) < p < (\hat{p} + E)$
$(0.6290 - 0.1202) < p < (0.6290 + 0.1202)$
$0.5088 < p < 0.7492$ or approximately 0.51 to 0.75.

We are 95% confident that the true proportion of actors who are extroverts is between 0.51 and 0.75, approximately. In repeated sampling from the same population, approximately 95% of the samples would generate confidence intervals that would enclose the true value of $\hat{p}$.

(c) $np \approx n\hat{p} = r = 39$
$nq \approx n\hat{q} = n - r = 62 - 39 = 23$
It is quite likely that np and $nq > 5$, since their estimates are much larger than 5. If np and $nq > 5$ then $\hat{p}$ is approximately normal with $\mu = p$ and $\sigma = \sqrt{pq/n}$. This forms the basis for the large sample confidence interval derivation.

3. $n = 5222,\ r = 1619$

(a) $\hat{p} = \frac{r}{n} = \frac{1619}{5222} = 0.3100$
so $\hat{q} = 1 - \hat{p} = 0.6900$

(b) $c = 99\%$, so $z_c = 2.58$
$E \approx z_c\sqrt{\hat{p}\hat{q}/n} = 2.58\sqrt{(0.3100)(0.6900)/5222} = 0.0165$
$(\hat{p} - E) < p < (\hat{p} + E)$
$(0.3100 - 0.0165) < p < (0.3100 + 0.0165)$
$0.2935 < p < 0.3265$ or approximately 0.29 and 0.33.

In repeated sampling, approximately 99% of the confidence intervals generated from the samples would include p, the proportion of judges who are hogans.

(c) $np \approx n\hat{p} = r = 1619, nq \approx n\hat{q} = n(1-\hat{p}) = n - r = 3603$
Since the estimates of np and nq are much greater than 5, it is reasonable to assume np and $nq > 5$. Then we can use the normal distribution with $\mu = p$ and $\sigma = \sqrt{pq/n}$ to approximate the distribution of $\hat{p}$.

5. $n = 5792,\ r = 3139$

(a) $\hat{p} = \frac{r}{n} = \frac{3139}{5792} = 0.5420$
so $\hat{q} = 1 - \hat{p} = 0.4580$

(b) $c = 99\%$, so $z_c = 2.58$

$E \approx z_c\sqrt{\hat{p}\hat{q}/n} = 2.58\sqrt{(0.5420)(0.4580)/5792} = 0.0169$

$(\hat{p} - E) < p < (\hat{p} + E)$

$(0.5420 - 0.0169) < p < (0.5420 + 0.0169)$

$0.5251 < p < 0.5589$, or approximately 0.53 to 0.56.

If we drew many samples of size 5792 physicians from those in Colorado, and generated a confidence interval from each sample, we would expect approximately 99% of the intervals to include the true proportion of Colorado physicians providing at least some charity care.

(c) $np \approx n\hat{p} = r = 3139 > 5; nq \approx n\hat{q} = n - r = 2653 > 5.$

Since the estimates of np and nq are much larger than 5, it is reasonable to assume np and nq are both greater than 5. Under the circumstances, it is appropriate to approximate the distribution of $\hat{p}$ with a normal distribution with $\mu = p$ and $\sigma = \sqrt{pq/n}$.

7. $n = 99, r = 17$

(a) $\hat{p} = \frac{r}{n} = \frac{17}{99} = 0.1717$

so $\hat{q} = 0.8283$

(b) $c = 85\%$, so $z_c = 1.44$

$E \approx z_c\sqrt{\hat{p}\hat{q}/n} = 1.44\sqrt{(0.1717)(0.8283)/99} = 0.0546$

$(\hat{p} - E) < p < (\hat{p} + E)$

$(0.1717 - 0.0546) < p < (0.1717 + 0.0546)$

$0.1171 < p < 0.2263$, or approximately 0.12 to 0.23.

In repeated sampling from this population about 85% of the intervals generated from these samples would include p, the proportion of fugitives arrested after their photographs appeared in the newspaper.

(c) $np \approx n\hat{p} = r = 17 > 5; nq \approx n\hat{q} = n - r = 82 > 5.$

Since the estimates of np and $nq > 5$, it is reasonable to assume np, $nq > 5$. If np and $nq > 5$, the normal distribution with $\mu = p$ and $\sigma = \sqrt{pq/n}$ provides a good approximation to the distribution of $\hat{p}$.

9. $n = 855, r = 26$

(a) $\hat{p} = \frac{r}{n} = \frac{26}{855} = 0.0304$

so $\hat{q} = 1 - \hat{p} = 0.9696$

(b) $c = 99\%$, so $z_c = 2.58$

$E \approx z_c\sqrt{\hat{p}\hat{q}/n} = 2.58\sqrt{(0.0304)(0.9696)/855} = 0.0151$

$(\hat{p} - E) < p < (\hat{p} + E)$

$(0.0304 - 0.0151) < p < (0.0304 + 0.0151)$

$0.0153 < p < 0.0455$, or approximately 0.02 to 0.05.

If many additional samples of size $n = 855$ were drawn from this fish population, and a confidence interval was created from each such sample, approximately 99% of those confidence intervals would contain p, the catch-and-release mortality rate (barbless hooks removed).

(c) $np \approx n\hat{p} = r = 26 > 5; nq \approx n\hat{q} = n - r = 829 > 5.$

Based on the estimates of np and nq, it is safe to assume both np and $nq > 5$. When np and $nq > 5$, the distribution of $\hat{p}$ can be accurately approximated by a normal distribution with $\mu = p$ and $\sigma = \sqrt{pq/n}$.

11. $n = 900$, $r = 54$, $n - r = 846$; both np and $nq > 5$.

(a) $\hat{p} = \frac{r}{n} = \frac{54}{900} = 0.0600$

so $\hat{q} = 1 - \hat{p} = 0.9400$

(b) $c = 99\%$, so $z_c = 2.58$

$E \approx z_c\sqrt{\hat{p}\hat{q}/n} = 2.58\sqrt{(0.0600)(0.9400)/900} = 0.0204$

$(\hat{p} - E) < p < (\hat{p} + E)$

$(0.0600 - 0.0204) < p < (0.0600 + 0.0204)$

$0.0396 < p < 0.0804$, or approximately 0.04 to 0.08.

13. $n = 2000$, $r = 382$, $n - r = 1618$; both np and $nq > 5$.

(a) $\hat{p} = \frac{r}{n} = \frac{382}{2000} = 0.1910$, so $\hat{q} = 0.8090$

(b) $c = 80\%$, so $z_c = 1.28$

$E \approx z_c\sqrt{\hat{p}\hat{q}/n} = 1.28\sqrt{(0.1910)(0.8090)/2000} = 0.0112$

$(\hat{p} - E) < p < (\hat{p} + E)$

$(0.1910 - 0.0112) < p < (0.1910 + 0.0112)$

$0.1798 < p < 0.2022$, or approximately 0.18 to 0.20.

15. $n = 730$, $r = 628$, $n - r = 102$; both np and $nq > 5$.

(a) $\hat{p} = \frac{r}{n} = \frac{628}{730} = 0.8603$, so $\hat{q} = 0.1397$

(b) $c = 95\%$, so $z_c = 1.96$

$E \approx z_c\sqrt{\hat{p}\hat{q}/n} = 1.96\sqrt{(0.8603)(0.1397)/730} = 0.0251$

$(\hat{p} - E) < p < (\hat{p} + E)$

$(0.8603 - 0.0251) < p < (0.8603 + 0.0251)$

$0.8352 < p < 0.8854$, or approximately 0.84 to 0.89.

In repeated sampling, approximately 95% of the intervals created from the samples would include p, the proportion of loyal women shoppers.

(c) Margin of error = $E \approx 2.5\%$

A recent study shows that about 86% of women shoppers remained loyal to their favorite supermarket last year. The study's margin of error is 2.5 percentage points.

17. $n = 1000$, $r = 250$, $n - r = 750$; both np and $nq > 5$.

(a) $\hat{p} = \frac{r}{n} = \frac{250}{1000} = 0.2500$, so $\hat{q} = 0.7500$

(b) $c = 95\%$, so $z_c = 1.96$

$E \approx z_c\sqrt{\hat{p}\hat{q}/n} = 1.96\sqrt{(0.2500)(0.7500)/1000} = 0.0268$

$(\hat{p} - E) < p < (\hat{p} + E)$

$(0.2500 - 0.0268) < p < (0.2500 + 0.0268)$

$0.2232 < p < 0.2768$, or approximately 0.22 to 0.28.

(c) Margin of error $= E \approx 2.7\%$

In a survey of 1000 large corporations, 25% admitted that, given a choice between equally qualified applicants, they would offer the job to the nonsmoker. The survey's margin of error was 2.7 percentage points.

Section 8.4

1. The goal is to estimate μ, the mean population of new lodgepole pine saplings in a 50-square-meter plot in Yellowstone National Park. Use $n = (z_c\sigma / E)^2$, where $c = 95\%$, $z_c = 1.96$, $\sigma = 44$, and $E = 10$.

$$n \approx \left[\frac{1.96(44)}{10}\right]^2 = 74.37, \text{ "round up" to 75 plots.}$$

3. The goal is to estimate the proportion, p, of people in your neighborhood who go to McDonald's. Use $n = pq(z_c/E)^2$.

(a) $c = 85\%, z_c = 1.44, E = 0.05$

With no preliminary estimate of p, we will use $p = q = 0.5$, which maximizes the value of $pq = p(1 - p)$, and, therefore, maximizes the value of n; no other choice of p would give a larger n. The "worst case" scenario, when pq and n are the largest values possible, occurs when $p = q = \frac{1}{2} = 0.5$.

Then, $n = pq(z_c / E)^2 \approx 0.5 \cdot 0.5(z_c / E)^2 = 0.25(z_c / E)^2 = 0.25(1.44 / 0.05)^2 = 207.36$, which is rounded up to 208 people.

(b) $\hat{p} = 8/90$ is a preliminary estimate of p; $q = 1 - p$, so $\hat{q} = 1 - \hat{p} = \dfrac{82}{90}$ is a preliminary estimate of q.

Then $n = pq(z_c / E)^2 \approx (8/90)(82/90)(1.44/0.05)^2 \approx 67.2$, or, rounded up, 68 people.

(Note that this sample size is much, much smaller than the sample size of $n = 208$ derived in part (a) where, in the absence of a preliminary estimate of p, we used the worst case estimate of $p, \hat{p} = 1/2$.)

5. The goal is to find μ, the mean player weight. Preliminary, $n = 56$.

Use $n = (z_c\sigma / E)^2$. $s = 26.58$, $c = 90\%$, $z_c = 1.645, E = 4$.

Then $n \approx [1.645(26.58)/4]^2 = 119.5$, or 120 players. However, since 56 players have already been drawn to estimate σ, we need only $120 - 56 = 64$ additional players.

7. The goal is to estimate the proportion of women students; use $n = \text{-}pq(z_c / E)^2$.

(a) $c = 99\%, z_c = 2.58, E = 0.05$

Since there is no preliminary estimate of p, we'll use the "worst case" estimate, $\hat{p} = \frac{1}{2}, \hat{q} = 1 - \hat{p} = \frac{1}{2}$.

$n \approx 0.5 \cdot 0.5(z_c / E)^2 = 0.25(z_c / E)^2 = 0.25(2.58/0.05)^2 = 665.6$, or 666 students.

(b) Preliminary estimate of p: $\hat{p} = 54\% = 0.54, \hat{q} = 1 - \hat{p} = 0.46$.

$n \approx (0.54)(0.46)(2.58/0.05)^2 = 661.4 \approx 662$ students.

(There is very little difference between (a) and (b) because (a) uses $\hat{p} = \frac{1}{2} = 0.50$ and (b) uses $\hat{p} = 0.54$, and the $\hat{p}s$ are approximately the same.)

9. The goal is to estimate μ, the mean reconstructed clay vessel diameter, use $n = (z_c\sigma / E)^2$.

Preliminary $n = 83$, $s = 5.5$, $c = 95\%$, $z_c = 1.96, E = 1.0$

$n \approx [1.96(5.5)/1.0]^2 = 116.2 \approx 117$ clay pots.

Since 83 pots were already measured, we need 117 – 83 = 34 additional reconstructed clay pots.

11. The goal is to estimate the proportion, p, of pine beetle-infested trees; use $n = pq(z_c / E)^2$.

(a) $c = 85\%, z_c = 1.44, E = 0.06$

No preliminary $\hat{p}$ available, so use "worst case" estimate of $p, \hat{p} = \frac{1}{2}; \hat{q} = 1 - \hat{p} = \frac{1}{2}$.

$n \approx 0.5 \cdot 0.5(z_c / E)^2 = 0.25(z_c / E)^2 = 0.25(1.44/0.06)^2 = 144$ trees.

(b) Preliminary study showed $r = 19$ infested trees in a sample of $n = 58$ trees, so $\hat{p} = r/n = 19/58 = 0.3276; \hat{q} = 1 - \hat{p} = 0.6724$.

$n = pq(z_c / E)^2 \approx (0.3276)(0.6724)(1.44/0.06)^2 = 126.88 \approx 127$ trees.

Since 58 trees have already been checked for infestation, we need to check 127 – 58 = 69 additional trees.

13. The goal is to estimate p, the proportion of voters favoring capital punishment; use $n = pq(z_c / E)^2$.

(a) $c = 99\%, z_c = 2.58, E = 0.01$

Since there is no preliminary estimate of p, we will use the "worst case" estimate, $\hat{p} = \frac{1}{2}; \hat{q} = 1 - \hat{p} = \frac{1}{2}$.

$n \approx 0.5 \cdot 0.5(z_c / E)^2 = 0.25(z_c / E)^2 = 0.25(2.58/0.01)^2 = 16{,}641$ people.

(b) Preliminary estimate $\hat{p} = 67\% = 0.67; \hat{q} = 1 - \hat{p} = 0.33$.

$n \approx (0.67)(0.33)(2.58/0.01)^2 = 14{,}717.3 \approx 14{,}718$ people.

(Comment: this is an exceptionally large sample size and it would cost David quite a large sum of money to survey this many people. Good surveys are often done with sample sizes of 300 to 1000. David needs to reconsider the confidence level, c, and the margin of error, E, he wants for his project. If David opted for $c = 95\%$, $z_c = 1.96$, and $E = 0.05$, he could use a sample of about 340.)

15. The goal is to estimate μ, the average time phone customers are on hold; use $n = (z_c\sigma / E)^2$.

Preliminary sample of size $n = 167$, $s = 3.8$ minutes, $c = 99\%$, $z_c = 2.58$, $E = 30$ seconds $= 0.5$ minute (all time figures must be in the same units).

$n \approx [2.58(3.8)/0.5]^2 = 384.5 \approx 385$ phone calls.

Since the airline already measured the time on hold for 167 calls, it needs to measure the time on hold for 385 – 167 = 218 more phone calls.

17. The goal is to estimate *p*, the proportion of pickup truck owners who are women; use $n = pq(z_c / E)^2$.

(a) $c = 90\%, z_c = 1.645, E = 0.1$

Since no preliminary estimate is available, use the worst case estimate: $\hat{p} = \frac{1}{2}; \hat{q} = 1 - \hat{p} = \frac{1}{2}$.

$n \approx 0.5 \cdot 0.5(z_c / E)^2 = 0.25(z_c / E)^2 = 0.25(1.645/0.1)^2 = 67.65 \approx 68$ owners.

(b) $\hat{p} = 24\% = 0.24; \hat{q} = 1 - \hat{p} = 0.76.$

$n \approx (0.24)(0.76)(1.645/0.1)^2 = 49.36 \approx 50$ owners.

19. (a) Hint: it is usually easier to go from the more complicated version to the easier version in showing the equality of two things. So, start with:

$$\begin{aligned} \frac{1}{4} - \left(p - \frac{1}{2}\right)^2 &= \frac{1}{4} - \left[p^2 - 2p\left(\frac{1}{2}\right) + \left(\frac{1}{2}\right)^2\right] \quad \text{Recall: } (a-b)^2 = a^2 - 2ab + b^2 \\ &= \frac{1}{4} - \left[p^2 - p + \frac{1}{4}\right] \\ &= -p^2 + p \\ &= p - p^2 = p(1-p) \end{aligned}$$

as was to be shown.

(b) For any number a, $a^2 \geq 0$, so

$$\left(p - \frac{1}{2}\right)^2 \geq 0$$

$$(-1)\left(p - \frac{1}{2}\right)^2 \leq (-1)(0) = 0 \quad \text{Multiply both sides by } -1 \text{, remembering to reverse the order of inequality.}$$

$$-\left(p - \frac{1}{2}\right)^2 \leq 0$$

$$0 \geq -\left(p - \frac{1}{2}\right)^2$$

$$\frac{1}{4} \geq \frac{1}{4} - \left(p - \frac{1}{2}\right)^2 \quad \text{Add } \tfrac{1}{4} \text{ to both sides.}$$

but $\frac{1}{4} - \left(p - \frac{1}{2}\right)^2 = p(1-p)$ from part (a), so $\frac{1}{4} \geq p(1-p)$

[$p(1-p)$ is never greater than $\frac{1}{4}$ because it is always less than or equal to $\frac{1}{4}$.]

21. The goal is to estimate the proportion, *p*, of votes for the Democratic presidential candidate. Use $n = pq(z_c / E)^2$.

(a) $E = 0.001, c = 99\%, z_c = 2.58$

Since there is no preliminary estimate of *p*, use $\hat{p} = \hat{q} = \frac{1}{2}$.

$n \approx 0.5 \cdot 0.5(z_c / E)^2 = 0.25(z_c / E)^2 = 0.25(2.58/0.001)^2 = 1{,}664{,}100$ votes.

(b) No. If the preliminary estimate of p was $\hat{p}=0.5$, the same as the no information, worst case estimate of p, the sample size would be exactly the same as in (a). The general formula $n = pq(z_c/E)^2$ with preliminary estimates of p (and q) = 0.5 is the same as the no-information-about-p formula, $n=\frac{1}{4}(z_c/E)^2$.

[The above solutions have repeatedly used this fact, demonstrating that using $\hat{p}=\hat{q}=0.5$ (the worst case estimates of p and q, giving the largest possible sample size that meets the stated criteria) in the special no-information-about-p formula, $n=\frac{1}{4}(z_c/E)^2$, directly.]

Section 8.5

Answers may vary slightly due to rounding.

1. (a) Using a calculator, the means and standard deviations round to the values given.

(b) $c = 90\%$, $d.f. = 12 - 1 = 11$ (smaller of n_1-1 and n_2-1), $t_c = 1.796$

$$E=t_c\sqrt{\frac{s_1^2}{n_1}-\frac{s_2^2}{n_2}}=1.796\sqrt{\frac{170.4^2}{12}+\frac{212.1^2}{16}}\approx 129.9$$

$$[(\bar{x}_1-\bar{x}_2)-E]<(\mu_1-\mu_2)<[(\bar{x}_1-\bar{x}_2)+E]$$
$$(747.5-738.9)-129.9<(\mu_1-\mu_2)<(747.5-738.9)+129.9$$
$$-121.3<(\mu_1-\mu_2)<138.5$$

(c) Because the interval contains both positive and negative numbers, we cannot say that one region is more interesting than the other at 90% confidence level.

(d) Student's t because σ_1 and σ_2 are unknown.

3. (a) Using a calculator, the means and standard deviations round to the values given.

(b) $c = 85\%$, $d.f. \approx 16 - 1 = 15$ (smaller of n_1-1 and n_2-1), $t_c = 1.517$

$$E=t_c\sqrt{\frac{s_1^2}{n_1}+\frac{s_2^2}{n_2}}=1.517\sqrt{\frac{7.93^2}{16}+\frac{12.26^2}{17}}\approx 5.42$$

$$[(\bar{x}_2-\bar{x}_2)-E]<(\mu_1-\mu_2)<[(\bar{x}_1-\bar{x}_2)+E]$$
$$[(51.66-33.60)-5.42]<(\mu_1-\mu_2)<[(51.66-33.60)+5.42]$$
$$12.64\%<(\mu_1-\mu_2)<23.48\%$$

(c) Because the interval contains only positive values, we can say at the 85% confidence level that technology companies receive a higher population mean percentage of foreign revenue.

(d) Use Student's t since σ_1 and σ_2 are unknown.

5. (a) Using a calculator, the means and standard deviations round to the values given.

(b) $c = 90\%$, $d.f. \approx 40 - 1 = 39$ (smaller of $n_1 - 1$ and $n_2 - 1$),
To use Table 6, round down to $d.f. = 35$, $t_c = 1.690$

$$E = t_c\sqrt{\frac{s_1^2}{n_1} + \frac{s_2^2}{n_2}} = 1.690\sqrt{\frac{0.366^2}{45} + \frac{0.314^2}{40}} \approx 0.125$$

$$[(\bar{x}_1 - \bar{x}_2) - E] < (\mu_1 - \mu_2) < [(\bar{x}_1 - \bar{x}_2) + E]$$
$$[(6.179 - 6.453) - 0.125] < (\mu_1 - \mu_2) < [(6.179 - 6.453) + 0.125]$$
$$-0.399 \text{ inch} < (\mu_1 - \mu_2) < -0.149 \text{ inch}$$

(c) Since the interval contains all negative numbers, it seems the population mean height of professional football players is less than that of professional basketball players at the 90% confidence level.

(d) Use Student's t since σ_1 and σ_2 are not known. Both samples are large, so no assumptions about the original distribution are needed.

7.

	Sample 1	Sample 2
n	375	571
r	289	23
$\hat{p}$	289/375 = 0.7707	23/571 = 0.0403

$n_1\hat{p}_1 = 289$, $n_1\hat{q}_1 = 86$, $n_2\hat{p}_2 = 23$, $n_2\hat{q}_2 = 548$
All four of these estimates are > 5.

(a) $c = 99\%$, $z_c = 2.58$

$$E \approx z_c\sqrt{\frac{\hat{p}_1\hat{q}_1}{n_1} + \frac{\hat{p}_2\hat{q}_2}{n_2}} = 2.58\sqrt{\frac{(0.7707)(0.2293)}{375} + \frac{(0.0403)(0.9597)}{571}} = 2.58(0.0232) = 0.0599 \approx 0.06$$

$[(\hat{p}_1 - \hat{p}_2) - E] < (p_1 - p_2) < [(\hat{p}_1 - \hat{p}_2) + E]$
$[(0.7707 - 0.0403) - 0.0599] < (p_1 - p_2) < [(0.7707 - 0.0403) + 0.0599]$
$0.6705 < (p_1 - p_2) < 0.7903$, or approximately 0.67 and 0.79

(b) Because the confidence interval contains only positive values, $p_1 > p_2$ at the 99% level.

9.

	Sample 1	Sample 2
n	9340	25,111
$\bar{x}$	63.3	72.1
σ	9.17	12.67

(a) $c = 99\%$, $z_c = 2.58$

$$E \approx z_c\sqrt{\frac{\sigma_1^2}{n_1} + \frac{\sigma_2^2}{n_2}} = 2.58\sqrt{\frac{9.17^2}{9340} + \frac{12.67^2}{25{,}111}} = 0.3201$$

$[(\bar{x}_1 - \bar{x}_2) - E] < (\mu_1 - \mu_2) < [(\bar{x}_1 - \bar{x}_2) + E]$
$[(63.3 - 72.1) - 0.3201] < (\mu_1 - \mu_2) < [(63.3 - 72.1) + 0.3201]$
$-9.1201 < (\mu_1 - \mu_2) < -8.4799$, or approximately -9.12 and -8.48

(b) The interval includes only negative numbers, leading us to believe $\mu_1 < \mu_2$ at the 99% level. The mean interval between Old Faithful eruptions during the period 1983 to 1987 is between 8.48 and 9.12 minutes longer than the mean interval between eruptions during the period 1948 to 1952. [Comment: it is highly unlikely the data in this problem constitute the required two independent random samples. First, the data are time series observations and are probably highly correlated. It is possible the 30 year gap between time periods would be sufficient to wipe out the effects of serial correlation so that the two samples could be considered independent. However, the times within each sample are still correlated, and random samples consist of data that are independent (and identically distributed). Second, the large sample sizes, much larger than needed, might indicate a census rather than a sample of data was used.]

11. $n_1 = 210, r_1 = 65, \hat{p}_1 = 65/210 = 0.3095, \hat{q}_1 = \dfrac{145}{210} = 0.6905$

$n_2 = 152, r_2 = 18, \hat{p}_2 = 18/152 = 0.1184, \hat{q}_2 = \dfrac{134}{152} = 0.8816.$

$n_1\hat{p}_1 = 65, n_1\hat{q}_1 = 145, n_2\hat{p}_2 = 18,$ and $n_2\hat{q}_2 = 134$ are all > 5.

(a) $c = 99\%, z_c = 2.58$

$$E \approx z_c\sqrt{\frac{\hat{p}_1\hat{q}_1}{n_1} + \frac{\hat{p}_2\hat{q}_2}{n_2}} = 2.58\sqrt{\frac{0.3095(0.6905)}{210} + \frac{(0.1184)(0.8816)}{152}} = 2.58(0.0413) = 0.1065$$

$$[(\hat{p}_1 - \hat{p}_2) - E] < (p_1 - p_2) < [(\hat{p}_1 - \hat{p}_2) + E]$$
$$[(0.3095 - 0.1184) - 0.1065] < (p_1 - p_2) < [(0.3095 - 0.1184) + 0.1065]$$
$$0.0846 < (p_1 - p_2) < 0.2976,$$ or approximately 0.085 to 0.298.

(b) The interval consists only of positive values, indicating $p_1 > p_2$. At the 99% confidence level, the difference in the percentage of traditional Navajo hogans is between 0.085 and 0.298, i.e., there are between 8.5% and 29.8% more hogans in the Fort Defiance Region than in the Indian Wells Region. If it is true that traditional Navajo tend to live in hogans, then, percentage-wise, there are more traditional Navajo at Fort Defiance than at Indian Hills.

13. **(a)** Using a calculator, the means and standard deviations round to the values given.

(b) $c = 85\%, d.f. \approx 10 - 1 = 9$ (smaller of $n_1 - 1$ and $n_2 - 1$), $t_c = 1.574$

$$E = t_c\sqrt{\frac{s_1^2}{n_1} + \frac{s_2^2}{n_2}} = 1.574\sqrt{\frac{8.32^2}{10} + \frac{8.87^2}{18}} \approx 5.3$$

$$[(\bar{x}_1 - \bar{x}_2) - E] < (\mu_1 - \mu_2) < [(\bar{x}_1 - \bar{x}_2) + E]$$
$$[(75.80 - 66.83) - 5.3] < (\mu_1 - \mu_2) < [(75.80 - 66.83) + 5.3]$$
$$3.67 \text{ pounds} < (\mu_1 - \mu_2) < 14.27 \text{ pounds}$$

or approximately 3.7 and 14.3 pounds.

(c) The confidence interval contains only positive values, so it appears that the population mean weight of grey wolves in the Chihuahua region is greater than that of grey wolves from the Durango region at the 85% confidence level.

15. Because the original group of 45 subjects was randomly split into 3 subgroups of 15, each subgrouping can be considered a random sample, and it is independent of the other subgroups/samples. We will use the Student's t distribution because σ_1 and σ_2 are not known. Since the sample sizes (15) are all < 30, the t distribution requires the data to be normal but is robust against some departures from normality, the authors of this study would have to make a case for their self esteem scores' distributions being at least mound-shaped (unimodal) and symmetric.

$c = 85\%, d.f. \approx 15 - 1 = 14, t_c = 1.523$

Preliminary calculations:

$\mu_i - \mu_j$	$\bar{x}_i - \bar{x}_j$	$E_{ij} = t_c\sqrt{\frac{s_i^2}{n_i} + \frac{s_j^2}{n_j}}$
1 *versus* 2	19.84 – 19.32 = 0.52	$1.523\sqrt{\frac{3.07^2}{15} + \frac{3.62^2}{15}} \approx 1.87$
1 *versus* 3	19.84 – 17.88 = 1.96	$1.523\sqrt{\frac{3.07^2}{15} + \frac{3.74^2}{15}} \approx 1.90$
2 *versus* 3	19.32 – 17.88 = 1.44	$1.523\sqrt{\frac{3.62^2}{15} + \frac{3.74^2}{15}} \approx 2.05$

$$[(\bar{x}_i - \bar{x}_j) - E_{ij}] < (\mu_i - \mu_j) < [(\bar{x}_i - \bar{x}_j) + E_{ij}]$$

(a) $i = 1, j = 2$:
$(0.52 - 1.87) < (\mu_1 - \mu_2) < (0.52 + 1.87)$
$-1.35 < (\mu_1 - \mu_2) < 2.39$

(b) $i = 1, j = 3$:
$(1.96 - 1.90) < (\mu_1 - \mu_3) < (1.96 + 1.90)$
$0.06 < (\mu_1 - \mu_3) < 3.86$

(c) $i = 2, j = 3$:
$(1.44 - 2.05) < (\mu_2 - \mu_3) < (1.44 + 2.05)$
$-0.61 < (\mu_2 - \mu_3) < 3.49$

(d) At the 85% level, we can say there is no significant difference between the self-esteem scores on competence and social acceptance, and no significant difference between self-esteem scores on social acceptance and physical attractiveness. However, the interval estimate for $\mu_1 - \mu_3$ contains only positive numbers, indicating that $\mu_1 > \mu_3$. At the 85% level we can say that the self-esteem score for competence was between 0.06 and 3.86 points higher than that for physical attractiveness.
Notes: (1) There are better ways to study these three self-esteem scores than the method used here; however, that technique is beyond the level of the text. (2) The confidence interval formula used in (a)-(c) is designed to capture the true difference, $\mu_i - \mu_j$, 85% of the time. However, the chance that all the intervals will simultaneously include the true value for their is and *is* is less than 85%. (For further details, consult a more advanced textbook and look up multiple comparisons, comparison-wise error rate, family-wise or experiment-wise error rate, or comparison wise, family-wise, and experiment wise confidence coefficients.) (3) Although (b) showed a statistically significant difference (at 85%) between μ_1 and μ_3, the paper's authors would have to argue whether a difference of 0.06 to 3.86 points has any practical significance. Statistical significance does not necessarily mean the results have practical significance.

17. (a) $[(\bar{x}_1 - \bar{x}_2) - E] < \mu_1 - \mu_2 < [(\bar{x}_1 - \bar{x}_2) + E]$

where $E \approx z_c \sqrt{\dfrac{s_1^2}{n_1} + \dfrac{s_2^2}{n_2}}$

$z_{0.90} = 1.645, z_{0.95} = 1.96, z_{0.99} = 2.58$

As we change from one c confidence interval to another, the only number that changes is z_c. The interval width is $2E$, and it depends on z_c; if c increases, z_c increases, and the interval gets wider; if c decreases, z_c decreases, and the interval gets narrower. The larger interval (larger z_c) always includes all the points that were in a smaller interval created from a smaller z_c. All intervals are centered at $(\bar{x}_1 - \bar{x}_2)$. Therefore, if a 95% confidence interval includes both positive and negative numbers, the 99% confidence interval must include both positive and negative numbers. However, in going from 95% to 90%, the width of the interval decreases evenly from both sides. If $(\bar{x}_1 - \bar{x}_2)$ is near 0, the 90% confidence interval could have both positive and negative values, just like the 95% confidence interval did, and all values in the 90% confidence interval would also have been in the 95% confidence interval. The farther $(\bar{x}_1 - \bar{x}_2)$ is from zero, however, the more likely it is that the narrower 90% confidence interval will contain only positive or only negative numbers.

(b) $[(\hat{p}_1 - \hat{p}_2) - E] < (p_1 - p_2) < [(\hat{p}_1 - \hat{p}_2) + E]$

where $E \approx z_c \sqrt{\dfrac{\hat{p}_1\hat{q}_1}{n_1} + \dfrac{\hat{p}_2\hat{q}_2}{n_2}}$

As in (a), the only number that changes (and, therefore, affects the width) is z_c, and all the intervals are centered at $(\hat{p}_1 - \hat{p}_2)$. If a 95% confidence interval contains only positive values, a 99% confidence interval could contain all positive numbers, or both positive and negative numbers, depending on how close $(\hat{p}_1 - \hat{p}_2)$ is to zero. However, with a narrower 90% confidence interval, if the 95% confidence interval had only positive values, the 90% interval will; every value in the 90% interval would also have been in the 95% confidence interval.

Note: the principle is the same whether it is a confidence interval for $\mu, \mu_1 - \mu_2, p, p_1 - p_2$, etc. One can make statements paralleling those of (b) for the 95% confidence interval having only negative values.

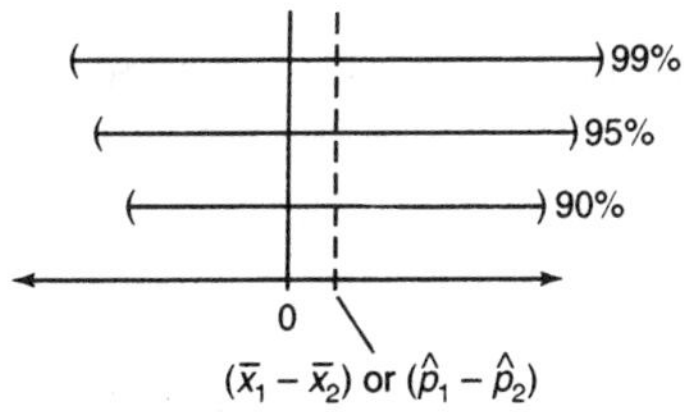

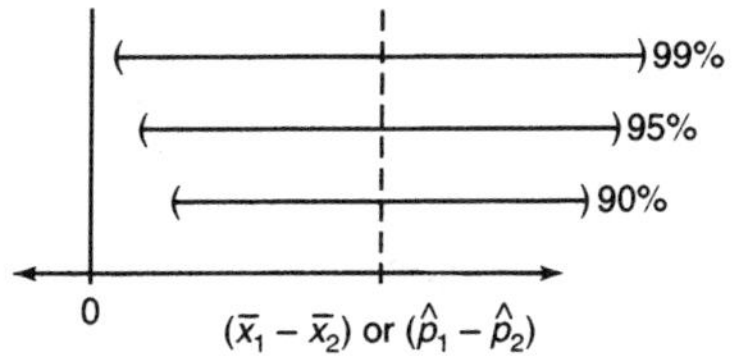

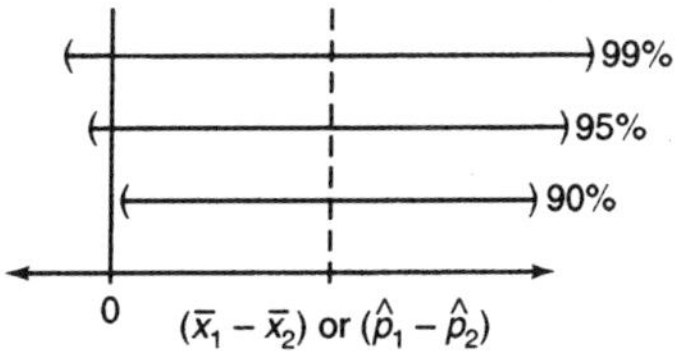

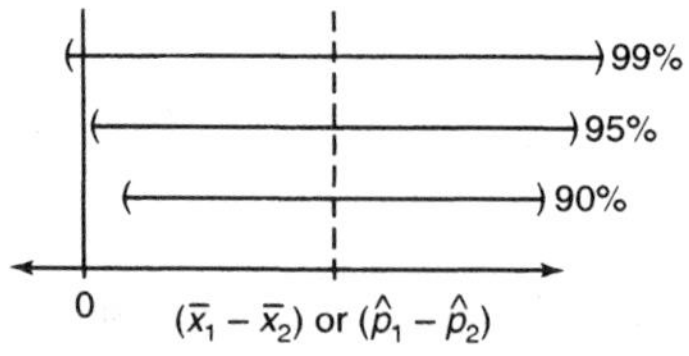

19. $E = z_c\sqrt{\dfrac{p_1q_1}{n_1} + \dfrac{p_2q_2}{n_2}} = \dfrac{z_c}{\sqrt{n}}\sqrt{p_1q_1 + p_2q_2}$ if $n = n_1 = n_2$

So $\sqrt{n}E = z_c\sqrt{p_1q_1 + p_2q_2}$ Multiply both sides by $\sqrt{n}$.

$\sqrt{n} = \dfrac{z_c}{E}\sqrt{p_1q_1 + p_2q_2}$ Divide both sides by E.

$n = \left(\dfrac{z_c}{E}\right)^2 (p_1q_1 + p_2q_2)$ Square both sides.

so $n \approx \left(\dfrac{z_c}{E}\right)^2 (\hat{p}_1\hat{q}_1 + \hat{p}_2\hat{q}_2)$

If we have no estimate for p_i, we would use the "worst case" estimate, i.e., the conservative approach would be to make n as large as possible, by using $\hat{p}_1 = 0.5$, $\hat{q}_1 = 1 - \hat{p}_1 = 0.5$

So, in this case,

$$n \approx \left(\frac{z_c}{E}\right)^2 [(0.5)(0.5) + (0.5)(0.5)] = 0.5\left(\frac{z_c}{E}\right)^2 \text{ or } \left(\frac{1}{2}\right)\left(\frac{z_c}{E}\right)^2$$

Again, all sample size estimates are the minimum number meeting the stated criteria.

(a) $c = 99\%, z_c = 2.58, E = 0.04,$

$\hat{p}_1 = \dfrac{289}{375} = 0.7707, \hat{q}_1 = 1 - \hat{p}_1 = 0.2293,$

$\hat{p}_2 = \dfrac{23}{571} = 0.0403, \hat{q}_2 = 0.9597$

(Recall the $n_i p_i$ and $n_i q_i$ conditions were checked in Problem 7 above.)

$$n \approx \left(\frac{z_c}{E}\right)^2 (\hat{p}_1\hat{q}_1 + \hat{p}_2\hat{q}_2) = \left(\frac{2.58}{0.04}\right)^2 [0.7707(0.2293) + 0.0403(0.9597)] = 896.107 \approx 897$$

n_1 and n_2 should each be 897 (married couples)

(b) Let $\hat{p}_i = \hat{q}_i = 0.5; c = 95\%, z_c = 1.96, E = 0.05$

$$n \approx \left(\frac{1}{2}\right)\left(\frac{z_c}{E}\right)^2 = \left(\frac{1}{2}\right)\left(\frac{1.96}{0.05}\right)^2 = 768.32 \approx 769$$

n_1 and n_2 should each be 769 (married couples)

21. **(a)** $c = 85\%,\ n_1 + n_2 - 2 = 10 + 18 - 2 = 26,\ t_{0.85}$ with 26 *d.f.* = 1.483

$$s_{\text{pooled}} = \sqrt{\frac{(n_1-1)s_1^2 + (n_2-1)s_2^2}{n_1+n_2-2}} = \sqrt{\frac{9(8.32^2)+17(8.87^2)}{10+18-2}} = 8.6836$$

$$E \approx t_c s_p \sqrt{\frac{1}{n_1}+\frac{1}{n_2}} = 1.483(8.6836)\sqrt{\frac{1}{10}+\frac{1}{18}} = 5.0791$$

$$[(\bar{x}_1 - \bar{x}_2) - E] < (\mu_1 - \mu_2) < [(\bar{x}_1 - \bar{x}_2) + E]$$
$$[(75.80 - 66.83) - 5.0791] < (\mu_1 - \mu_2) < [75.80 - 66.83) + 5.0791]$$
$$3.9 < (\mu_1 - \mu_2) < 14.1$$

(b) The pooled standard deviation method has a shorter interval and larger degrees of freedom.

Chapter 8 Review

1. **point estimate:** a single number used to estimate a population parameter
critical value: the x-axis values (arguments) of a probability density function (such as the standard normal or Student's t) which cut off an area of $c, 0 \le c \le 1,$ under the curve between them. Examples: the area under the standard normal curve between $-z_c$ and $+z_c$ is c; the area under the curve of a Student's t distribution between $-t_c$ and $+t_c$ as c. The area is symmetric about the curve's mean, μ.
maximal margin of error, *E*: the largest distance ("error") between the point estimate and the parameter it estimates that can be tolerated under certain circumstances; E is the half-width of a confidence interval.
confidence level, *c*: A measure of the reliability of an (interval) estimate: c denotes the proportion of all possible confidence interval estimates of a parameter (or difference between 2 parameters) that will include the true value being estimated. It is a statement about the probability the procedure being used will capture the value of interest; it cannot be considered a measure of the reliability of a specific interval, because any specific interval is either right or wrong—either it captures the parameter value, or it does not.
confidence interval: a procedure designed to give a range of values as an (interval) estimate of an unknown parameter value; compare point estimate. What separates confidence interval estimates from any other interval estimate (such as 4, give or take 2.8) is that the reliability of the procedure can be determined: if $c = 0.90 = 90\%$, for example, a 90% confidence interval (estimate) for μ says that if all possible samples of size n were drawn, and a 90% confidence interval for μ was created for each such sample using the prescribed method (such as $\bar{x} \pm z_c\sigma/\sqrt{n}$), then if the true value of μ became known, 90% of the confidence intervals so created would include the value of μ.

3. $n = 73,\ \bar{x} = 178.70,\ \sigma = 7.81,\ c = 95\%,\ z_c = 1.96$

$$E \approx \frac{z_c\sigma}{\sqrt{n}} = \frac{1.96(7.81)}{\sqrt{73}} = 1.7916 \approx 1.79$$
$$(\bar{x} - E) < \mu < (\bar{x} + E)$$
$$(178.70 - 1.79) < \mu < (178.70 + 1.79)$$
$$176.91 < \mu < 180.49$$

5. **(a)** $\bar{x} = 74.2, s = 18.2530 \approx 18.3$, as indicated

(b) $c = 95\%, n = 15,\ t_{0.95}$ with $n - 1 = 14$ *d.f.* = 2.145

$$E \approx \frac{t_c s}{\sqrt{n}} = \frac{2.145(18.3)}{\sqrt{15}} = 10.1352 \approx 10.1$$
$$(\bar{x} - E) < \mu < (\bar{x} + E)$$
$$(74.2 - 10.1) < \mu < (74.2 + 10.1)$$
$$64.1 \text{ centimeters} < \mu < 84.3 \text{ centimeters}$$

7. $n = 2958,\ r = 1538,\ \hat{p} = \frac{r}{n} = \frac{1538}{2958} = 0.5199 \approx 0.52$

$\hat{q} = 1 - \hat{p} = 0.4801,\ n - r = 1420,\ c = 90\%,\ z_c = 1.645$

$np \approx n\hat{p} = r = 1538 > 5,\ nq \approx n\hat{q} = n - r = 1420 > 5$

$E \approx z_c\sqrt{\frac{\hat{p}\hat{q}}{n}} = 1.645\sqrt{\frac{(0.5199)(0.4801)}{2958}} = 0.0151 \approx 0.02$

$(\hat{p} - E) < p < (\hat{p} + E)$

$(0.52 - 0.02) < p < (0.52 + 0.02)$

$0.50 < p < 0.54$

9. $n = 167, r = 68$

(a) $\hat{p} = \frac{r}{n} = \frac{68}{167} = 0.4072, \hat{q} = 0.5928, n - r = 99$

(b) $n\hat{p} = r = 68 > 5$ and $n\hat{q} = n - r = 99 > 5$

$c = 95\%, z_c = 1.96$

$E \approx z_c\sqrt{\frac{\hat{p}\hat{q}}{n}} = 1.96\sqrt{\frac{(0.4072)(0.5928)}{167}} = 0.0745$

$(\hat{p} - E) < p < (\hat{p} + E)$

$(0.4072 - 0.0745) < p < (0.4072 + 0.0745)$

$0.3327 < p < 0.4817$, or approximately 0.333 to 0.482

11. **(a)** Using a calculator, the means and standard deviations round to the values given.

(b) $c = 95\%$, *d.f.* is smaller of $n_1 - 1$ and $n_2 - 1$, $d.f. \approx n_1 - 1 = 72 - 1 = 71$ (round down to 70 for Table 6), $t_c = 1.994$

$E = t_c\sqrt{\frac{s_1^2}{n_1} + \frac{s_2^2}{n_2}} = 1.994\sqrt{\frac{2.08^2}{72} + \frac{3.03^2}{80}} \approx 0.83$

$[(\bar{x}_1 - \bar{x}_2) - E] < (\mu_1 - \mu_2) < [(\bar{x}_1 - \bar{x}_2) + E]$

$[(11.42 - 10.65) - 0.83] < (\mu_1 - \mu_2) < [(11.42 - 10.65) + 0.83]$

$-0.06 < (\mu_1 - \mu_2) < 1.6$

(c) Because the interval contains both positive and negative values, we cannot conclude there is any difference in soil water content in the two fields at the 95% confidence level.

(d) Student's *t* distribution because σ_1 and σ_2 are not known. Both samples are large, so no assumptions about the original distribution are needed.

13. $n_1 = 18,\ \bar{x}_1 = 98,\ s_1 = 6.5,\ n_2 = 24,\ \bar{x}_2 = 90,\ s_2 = 7.3$

(a) $c = 75\%$, $d.f. \approx 18 - 1 = 17$ (smaller of $n_1 - 1$ and $n_2 - 1$),
$t_c = 1.191$

$$E = t_c\sqrt{\frac{s_1^2}{n_1} + \frac{s_2^2}{n_2}} = 1.191\sqrt{\frac{6.5^2}{18} + \frac{7.3^2}{24}} \approx 2.5$$

$$[(\bar{x}_1 - \bar{x}_2) - E] < (\mu_1 - \mu_2) < [(\bar{x}_1 - \bar{x}_2) + E]$$
$$[(98 - 90) - 2.5] < (\mu_1 - \mu_2) < [(98 - 90) + 2.5]$$
$$5.5 \text{ pounds} < (\mu_1 - \mu_2) < 10.5 \text{ pounds}$$

(b) Since the interval contains only positive values, we can say that $\mu_1 > \mu_2$ at the 75% level, i.e., Canadian wolves weigh more than Alaska wolves, and that the difference is approximately 5.5 to 10.5 pounds.

15. $n_1 = 93,\ r_1 = 79,\ \hat{p}_1 = \frac{79}{93} = 0.8495,\ \hat{q}_1 = 0.1505,\ n_1 - r_1 = 14$

$n_2 = 83,\ r_2 = 74,\ \hat{p}_2 = \frac{74}{83} = 0.8916,\ \hat{q}_2 = 0.1084,\ n_2 - r_2 = 9$

$n_i\hat{p}_i$ and $n_i\hat{q}_i$ are all > 5

(a) $c = 95\%, z_c = 1.96$

$$E \approx z_c\sqrt{\frac{\hat{p}_1\hat{q}_1}{n_1} + \frac{\hat{p}_2\hat{q}_2}{n_2}} = 1.96\sqrt{\frac{(0.8495)(0.1505)}{93} + \frac{(0.8916)(0.1084)}{83}} = 0.0988$$

$$[(\hat{p}_1 - \hat{p}_2) - E] < (p_1 - p_2) < [(\hat{p}_1 - \hat{p}_2) + E]$$
$$[(0.8495 - 0.8916) - 0.0988] < (p_1 - p_2) < [(0.8495 - 0.8916) + 0.0988]$$
$$-0.1409 < (p_1 - p_2) < 0.0567$$

(b) Since the interval contains positive and negative values, we can say that at the 95% level, there is no significant difference between the proportion of accurate responses to face-to-face interviews and that of accurate responses to telephone interviews.

17. (a) $P(A_1 < \mu_1 < B_1) = 0.80$
$P(A_2 < \mu_2 < B_2) = 0.80$
i.e., the two intervals are designed so that the confidence interval procedure produces intervals A_i to B_i that include μ_i 80% of the time.

$P(A_1 < \mu_1 < B_1 \text{ and } A_2 < \mu_2 < B_2) = P(A_1 < \mu_1 < B_1) \cdot P(A_2 < \mu_2 < B_2, \text{ given } A_1 < \mu_1 < B_1)$

but the intervals were created using independent samples, so the intervals themselves are independent, so

$= P(A_1 < \mu_1 < B_1) \cdot P(A_2 < \mu_2 < B_2)$

by the definition of independent events:
if *C* and *D* are independent, *P(C,* given *D)* = *P(C)*.

$= (0.80)(0.80)$

$= 0.64$

The probability that both intervals are simultaneously correct, i.e., that both intervals capture their μ_i, is 0.64.

$P(\text{at least one interval fails to capture its } \mu_i) = 1 - P(\text{both intervals capture their } \mu_1)$
$= 1 - 0.64$
$= 0.36$

[There are 4 possible outcomes. Using the (*x*, *y*) interval notation, they are

(1) $\mu_1 \in (A_1, B_1)$, $\mu_2 \in (A_2, B_2)$ both capture μ_i

(2) $\mu_1 \in (A_1, B_1)$, $\mu_2 \notin (A_2, B_2)$ only μ_1, is captured

(3) $\mu_1 \notin (A_1, B_1)$, $\mu_2 \in (A_2, B_2)$ only μ_2 is captured

(4) $\mu_1 \notin (A_1, B_1)$, $\mu_2 \notin (A_2, B_2)$ neither μ_i is captured

By the complimentary event rule, *P*(at least 1 fails) = *P*[case (2), (3), or (4)] = 1 – *P*[case (1)]

(b) $P(A_1 < \mu_1 < B_1) = c$
$P(A_2 < \mu_2 < B_2) = c$

(Both confidence intervals are at level *c*.)

$0.90 = P(A_1 < \mu_1 < B_1 \text{ and } A_2 < \mu_2 < B_2)$
$= P(A_1 < \mu_1 < B_1) \bullet P(A_2 < \mu_2 < B_2)$ since the intervals are independent
$= c \bullet c$
$= c^2$

If $0.90 = c^2$, then $\sqrt{0.90} = c = 0.9487$, or approximately 0.95.

(c) Answers vary.
In large, complex engineering designs, each component must be within design specifications or the project will fail.
Consider the hundreds (if not thousands) of components which must function properly to launch the space shuttle, keep it orbiting, and return it safely to earth. For example, nuts, bolts, rivets, wiring, and the like must be a certain size, give or take some tiny amount. Tiles and the glue securing them must be able to withstand a huge range of temperatures, from the ambient air temperature at launch time to the extreme heat of re-entry.
Each of the design specifications can be thought of as a confidence interval. Manufacturers and suppliers want to be very confident their parts are well within the specifications, or they might lose their contracts to competitors. Similarly, NASA wants to be completely confident that all the parts meet specifications as a whole; otherwise, costly delays or catastrophic failures may occur. (Recall that much of the Challenger disaster was due to O-ring failure-because NASA decided to go ahead with the launch even though they had been warned by the O-ring manufacturer that the temperature at the launch site was below the lowest temperature at which the O-rings had been tested, and that the O-rings might not completely seat at that temperature.)
If NASA will tolerate only a 1 in 1,000 or 1 in 1,000,000 chance of failure, i.e., $c = 0.999$ or $c = 0.999999$, the individual components' confidence levels, *c* must be (much) higher than NASA's.

Chapter 9 Hypothesis Testing

Section 9.1

1. See text for definitions. Essay may include:

(a) A working hypothesis about the population parameter in question is called the null hypothesis. The value specified in the null hypothesis is often a historical value, a claim, or a production specification.

(b) Any hypothesis that differs from the null hypothesis is called an alternate hypothesis.

(c) If we reject the null hypothesis when it is in fact true, we have an error that is called a type I error. On the other hand, if we accept (i.e., fail to reject) the null hypothesis when it is in fact false, we have made an error that is called a type II error.

(d) The probability with which we are willing to risk a type I error is called the level of significance of a test. The probability of making a type II error is denoted by β.

3. No, if we fail to reject the null hypothesis, we have not proven it to be true beyond all doubt. The evidence is not sufficient to merit rejecting H_0.

5. (a) The claim is $\mu = 60$ kg, so you would use $H_0: \mu = 60$ kg.

(b) We want to know if the average weight is less than 60 kg, so you would use $H_1: \mu < 60$ kg.

(c) We want to know if the average weight is greater than 60 kg, so you would use $H_1: \mu > 60$ kg.

(d) We want to know if the average weight is different from (more or less than) 60 kg, so you would use $H_1: \mu \neq 60$ kg.

(e) Since part (b) is a left-tailed test, the critical region is on the left. Since part (c) is a right-tailed test, the critical region is on the right. Since part (d) is a two-tailed test, the critical region is on both sides of the mean.

7. (a) The claim is $\mu = 16.4$ ft, so $H_0: \mu = 16.4$ ft.

(b) You want to know if the average is getting larger, so $H_1: \mu > 16.4$ ft.

(c) You want to know if the average is getting smaller, so $H_1: \mu < 16.4$ ft.

(d) You want to know if the average is different from 16.4 ft, so $H_1: \mu \neq 16.4$ ft.

(e) Since part (b) is a right-tailed test, the area corresponding to the P-value is on the right. Since part (c) is a left-tailed test, the area corresponding to the P-value is on the left. Since part (d) is a two-tailed test, the area corresponding to the P-value is on both sides of the mean.

9. (a) $\alpha = 0.01$
$H_0: \mu = 4.7\%$
$H_1: \mu > 4.7\%$
Since > is in H_1, use a right-tailed test.

(b) Use the standard normal distribution. We assume x has a normal distribution with known standard deviation σ. Note that μ is given in the null hypothesis.

$$z = \frac{\bar{x} - \mu}{\frac{\sigma}{\sqrt{n}}} = \frac{5.38 - 4.7}{\frac{2.4}{\sqrt{10}}} \approx 0.90$$

(c) P-value $= P(z > 0.90) = 0.1841$

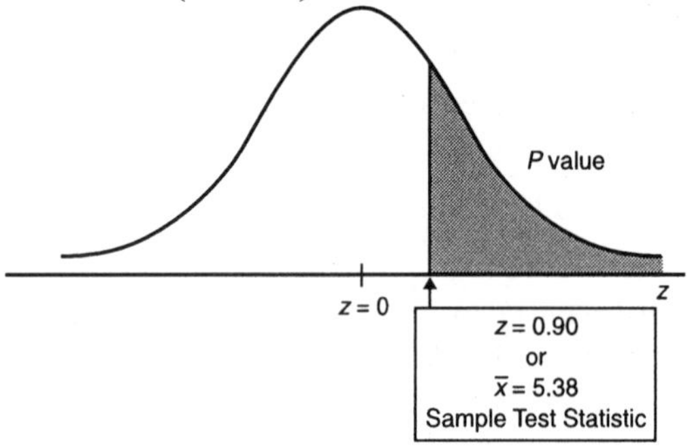

(d) Since a P-value of $0.1841 > 0.01$ for α, we fail to reject H_0. No.

(e) There is insufficient evidence at the 0.01 level to reject the claim that average yield for bank stocks equals average yield for all stocks.

11. (a) $\alpha = 0.01$
H_0: $\mu = 4.55$ g
H_1: $\mu < 4.55$ g
Since $<$ is in H_1, use a left-tailed test.

(b) Use the standard normal distribution. We assume x has a normal distribution with known standard deviation σ. Note that μ is given in the null hypothesis.

$$z = \frac{\bar{x} - \mu}{\frac{\sigma}{\sqrt{n}}} = \frac{3.75 - 4.55}{\frac{0.70}{\sqrt{6}}} \approx -2.80$$

(c) P-value $= P(z < -2.80) = 0.0026$

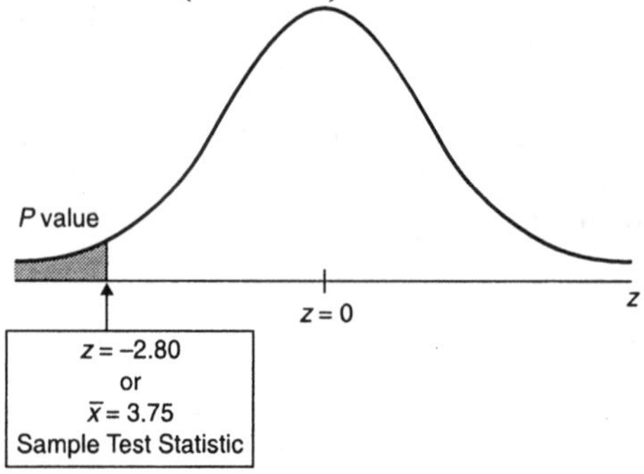

(d) Since a P-value of $0.0026 \le 0.01$ for α, we reject H_0. Yes.

(e) The sample evidence is sufficient at the 0.01 level to justify rejecting H_0. It seems the humming birds in the Grand Canyon region have a lower average weight.

13. (a) $\alpha = 0.01$
H_0: $\mu = 11\%$
H_1: $\mu \neq 11\%$
Since $\neq$ is in H_1, use a two-tailed test.

(b) Use the standard normal distribution. We assume x has a normal distribution with known standard deviation σ. Note that μ is given in the null hypothesis.

$$z = \frac{\bar{x} - \mu}{\frac{\sigma}{\sqrt{n}}} = \frac{12.5 - 11}{\frac{5.0}{\sqrt{16}}} = 1.20$$

(c) P-value = $2P(z > 1.20) = 2(0.1151) = 0.2302$

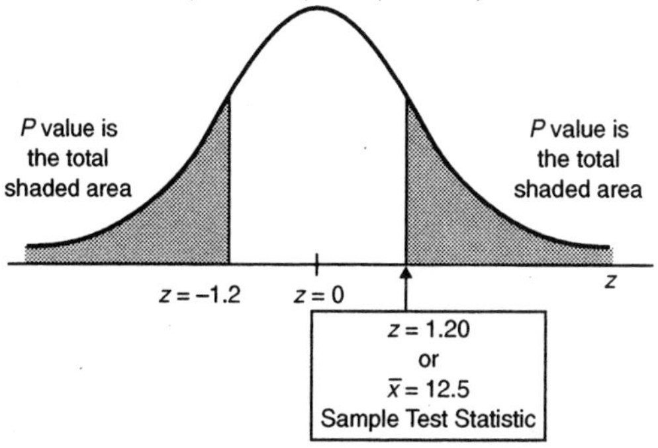

(d) Since a P-value of $0.2302 > 0.01$ for α, we fail to reject H_0. No.

(e) There is insufficient evidence at the 0.01 level of significance to reject H_0. It seems the average hail damage to wheat crops in Weld County matches the national average.

Section 9.2

1. (a) $\alpha = 0.01$
H_0: $\mu = 16.4$ ft
H_1: $\mu > 16.4$ ft

(b) Use the standard normal distribution. The sample size is large, $n \geq 30$, and we know σ.

$$z = \frac{\bar{x} - \mu}{\frac{\sigma}{\sqrt{n}}} = \frac{17.3 - 16.4}{\frac{3.5}{\sqrt{36}}} \approx 1.54$$

(c) P-value = $P(z > 1.54) = 0.0618$

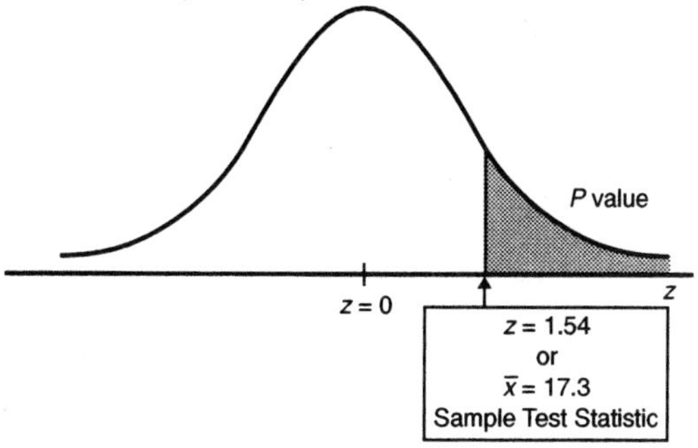

(d) Since a P-value of $0.0618 > 0.01$ for α, we fail to reject H_0. No.

(e) At the 1% level of significance, there is insufficient evidence to say average storm level is increasing.

3. (a) $\alpha = 0.05$
$H_0: \mu = 41.7$
$H_1: \mu \neq 41.7$

(b) Use a standard normal distribution. The sample size is large, $n \geq 30$, and we know σ.

$$z = \frac{\bar{x} - \mu}{\frac{\sigma}{\sqrt{n}}} = \frac{36.2 - 41.7}{\frac{18.5}{\sqrt{45}}} \approx -1.99$$

(c) P-value $= 2P(z < -1.99) = 2(0.0233) = 0.0466$

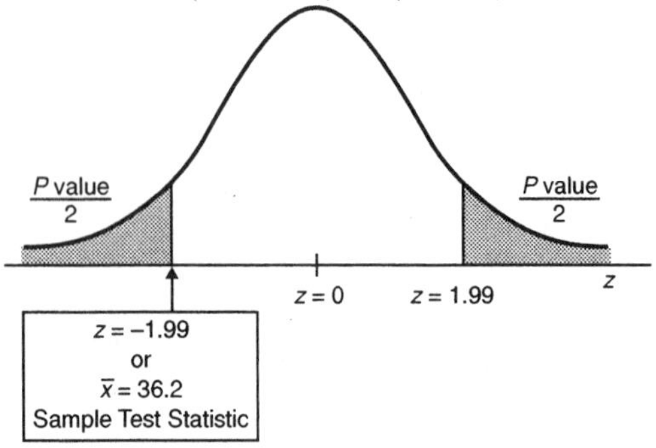

(d) Since a P-value of $0.0466 \leq 0.05$ for α, we reject H_0. Yes.

(e) At the 5% level of significance, there is sufficient evidence to say the average number of e-mails is different with the new priority system.

5. (a) $\alpha = 0.01$
$H_0: \mu = 1.75$ yr
$H_1: \mu > 1.75$ yr

(b) Use the Student's t distribution with $d.f. = n - 1 = 46 - 1 = 45$. The sample size is large, $n \geq 30$, and σ is unknown.

$$t = \frac{\bar{x} - \mu}{\frac{s}{\sqrt{n}}} = \frac{2.05 - 1.75}{\frac{0.82}{\sqrt{46}}} \approx 2.481$$

(c) For $d.f. = 45$, 2.481 falls between entries 2.412 and 2.690. Use one-tailed areas to find that $0.005 < P\text{-value} < 0.010$.

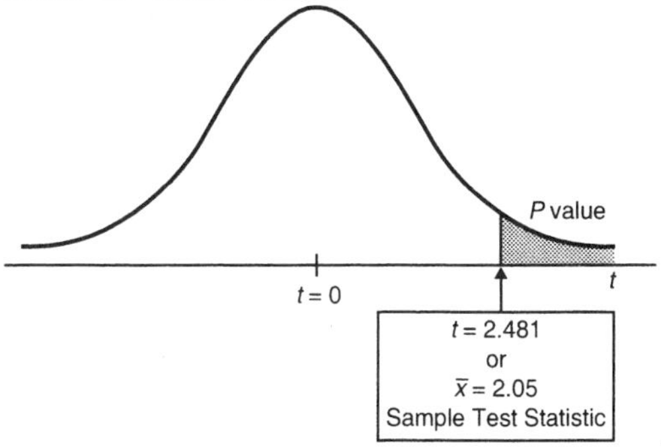

(d) Since the entire P-value interval ≤ 0.01 for α, we reject H_0. Yes.

(e) At the 1% level of significance, sample data indicate the average age of the Minnesota region Coyotes is higher than 1.75 years.

7. (a) $\alpha = 0.05$
H_0: $\mu = 19.4$
H_1: $\mu \neq 19.4$

(b) Use the Student's t distribution with $d.f. = n - 1 = 36 - 1 = 35$. The sample size is large, $n \geq 30$, and σ is unknown.

$$t = \frac{\bar{x} - \mu}{\frac{s}{\sqrt{n}}} = \frac{17.9 - 19.4}{\frac{5.2}{\sqrt{36}}} \approx -1.731$$

(c) For $d.f. = 35$, 1.731 falls between entries 1.690 and 2.030. Use two-tailed areas to find that $0.05 < P\text{-value} < 0.100$.

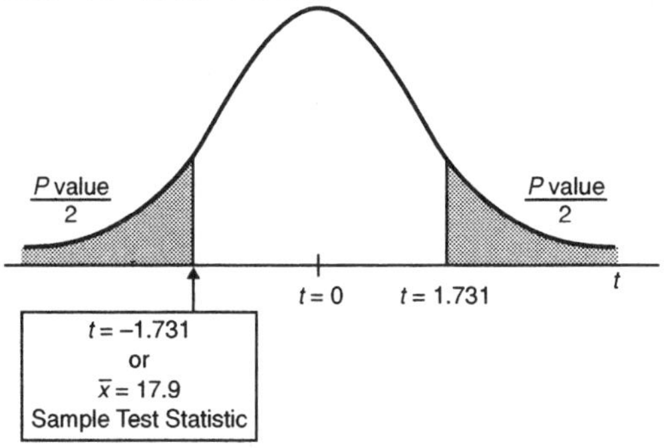

(d) Since the P-value interval > 0.05 for α, we fail to reject H_0. No.

(e) At the 5% level of significance, the sample evidence does not support rejecting the claim that the average P/E for socially responsible funds is different from that of the S&P Stock Index.

9. (i) Use a calculator to verify. Rounded values are used in part ii.

(ii) (a) $\alpha = 0.05$
H_0: $\mu = 4.8$
H_1: $\mu < 4.8$

(b) Use the Student's t distribution with $d.f. = n - 1 = 6 - 1 = 5$. We assume x has a distribution that is approximately normal and σ is unknown.

$$t = \frac{\bar{x}-\mu}{\frac{s}{\sqrt{n}}} = \frac{4.40-4.8}{\frac{0.28}{\sqrt{6}}} \approx -3.499$$

(c) For $d.f. = 5$, 3.499 falls between entries 3.365 and 4.032. Use one-tailed areas to find that $0.005 < P\text{-value} < 0.010$.

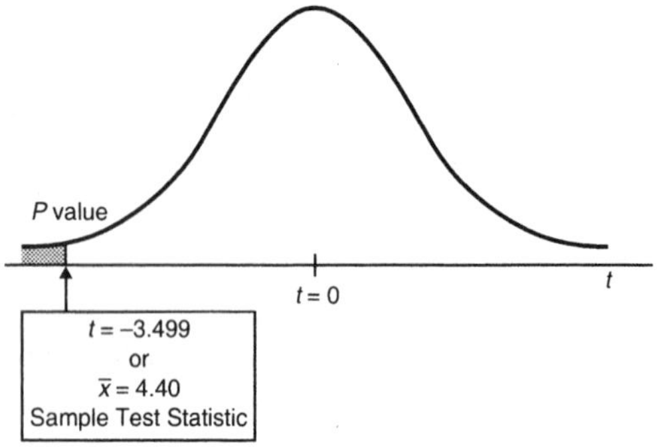

(d) Since the entire P-value interval ≤ 0.05 for α, we reject H_0. Yes.

(e) At the 5% level of significance, sample evidence supports the claim that the average RBC count for this patient is less than 4.8.

11. (i) Use a calculator to verify. Rounded values are used in part ii.

(ii) (a) $\alpha = 0.01$

H_0: $\mu = 67$

H_1: $\mu \ne 67$

(b) Use the Student's t distribution with $d.f. = n - 1 = 16 - 1 = 15$. We assume x has a distribution that is approximately normal and σ is unknown.

$$t = \frac{\bar{x}-\mu}{\frac{s}{\sqrt{n}}} = \frac{61.8-67}{\frac{10.6}{\sqrt{16}}} \approx -1.962$$

(c) For $d.f. = 15$, 1.962 falls between entries 1.753 and 2.131. Use two-tailed areas to find that $0.050 < P\text{-value} < 0.100$.

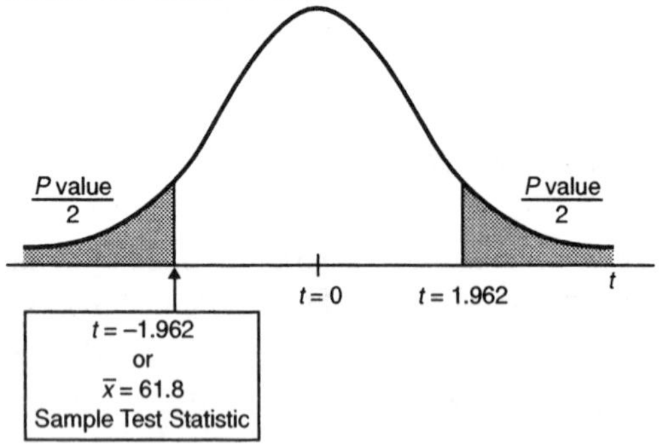

(d) Since P-value interval > 0.01 for α, we fail to reject H_0. No.

(e) At the 1% level of significance, sample evidence does not support a claim that the average thickness of slab avalanches in Vail is different from those in Canada.

13. (i) Use a calculator to verify. Rounded values are used in part ii.

(ii) (a) $\alpha = 0.05$
H_0: $\mu = 40$ beats per min
H_1: $\mu \neq 40$ beats per min

(b) Use the Student's t distribution with $d.f. = n - 1 = 6 - 1 = 5$. We assume x has a distribution that is approximately normal and σ is unknown.

$$t = \frac{\bar{x} - \mu}{\frac{s}{\sqrt{n}}} = \frac{36.5 - 40}{\frac{4.2}{\sqrt{6}}} \approx -2.041$$

(c) For $d.f. = 5$, 2.041 falls between entries 2.015 and 2.571. Use two-tailed areas to find that $0.050 < P\text{-value} < 0.100$.

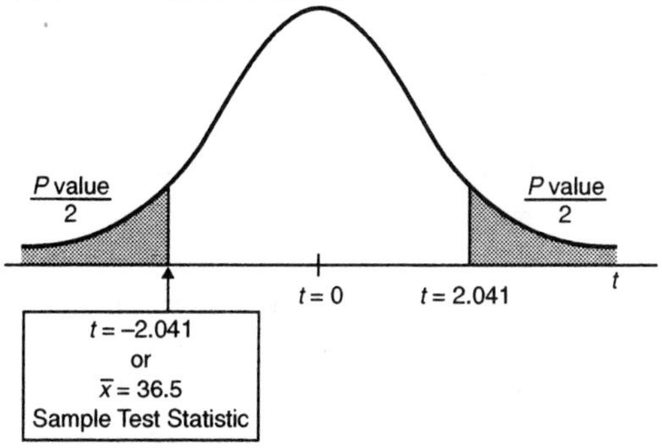

(d) Since the entire P-value interval > 0.05 for α, we fail to reject H_0. No.

(e) At the 5% level of significance, the population average heart rate for the lion is not significantly different.

15. (i) Use a calculator to verify. Rounded values are used in part ii.

(ii) (a) $\alpha = 0.05$
H_0: $\mu = 8.8$
H_1: $\mu \neq 8.8$

(b) Use the Student's t distribution with $d.f. = n - 1 = 14 - 1 = 13$. We assume x has a distribution that is approximately normal and σ is unknown.

$$t = \frac{\bar{x} - \mu}{\frac{s}{\sqrt{n}}} = \frac{7.36 - 8.8}{\frac{4.03}{\sqrt{14}}} \approx -1.337$$

(c) For *d.f.* = 13, 1.337 falls between entries 1.204 and 1.350. Use two-tailed areas to find that $0.200 < P\text{-value} < 0.250$.

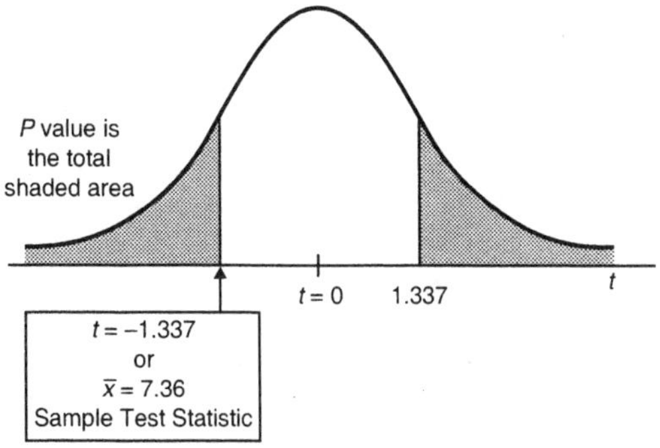

(d) Since *P*-value interval > 0.05 for α, we fail to reject H_0. No.

(e) At the 5% level of significance, we cannot conclude the catch is different from 8.8 fish per day.

17. (a) *P*-value of a one-tailed test is smaller. For a two-tailed test, the *P*-value is double because it includes the area in both tails.

(b) Yes; the *P*-value of a one-tailed test is smaller, so it might be smaller than α while the *P*-value of a two-tailed test is larger than α.

(c) Yes; if the two-tailed *P*-value is less than α, the smaller one-tail area is also less than α.

(d) Yes; the conclusions can be different. The conclusion based on the two-tailed test is more conservative in the sense that the sample data must be more extreme (differ more from H_0) in order to reject H_0.

19. (a) H_0: $\mu = 20$

H_1: $\mu \neq 20$

For $\alpha = 0.01$, $c = 1 - 0.01 = 0.99$, $\sigma = 4$, and $z_c = 2.58$.

$$E \approx z_c \frac{\sigma}{\sqrt{n}} = 2.58 \frac{4}{\sqrt{36}} = 1.72$$

$$\bar{x} - E < \mu < \bar{x} + E$$

$$22 - 1.72 < \mu < 22 + 1.72$$

$$20.28 < \mu < 23.72$$

The hypothesized mean $\mu = 20$ is not in the interval. Therefore, we reject H_0.

(b) Because $n = 36$ is large, the sampling distribution of $\bar{x}$ is approximately normal by the central limit theorem, and we know σ.

$$z = \frac{\bar{x} - \mu}{\sigma/\sqrt{n}} = \frac{22 - 20}{4/\sqrt{36}} = 3.00$$

From Table 5, *P*-value = $2P(z < -3.00) = 2(0.0013) = 0.0026$. Since $0.0026 \leq 0.01$ for α, we reject H_0. The results are the same.

21. For a right-tailed test and $\alpha = 0.01$, the critical value $z_0 = 2.33$; critical region is values to the right of 2.33. Since the sample statistic $z = 1.54$ is not in the critical region, we fail to reject H_0. At the 1% level, there is insufficient evidence to say the average storm level is increasing. Conclusion is same as with P-value method.

23. For a two-tailed test and $\alpha = 0.05$, the critical value $z_0 = \pm 1.96$; critical regions are values to the left of -1.96 together with values to the right of 1.96. Since the sample test statistic $z = -1.99$ is in the critical region, we reject H_0. At the 5% level, there is sufficient evidence to say the average number of e-mails is different with the new priority system. Conclusions is same as with P-value method.

25. For a right-tailed test and $\alpha = 0.01$, critical value is $t_0 = 2.412$ with $d.f. = 45$. Critical region is values to the right of 2.412. Since the sample test statistic $t = 2.481$ is in the critical region, we reject H_0. At the 1% level sample data indicate the average age of Minnesota Region Coyotes is higher than 1.75 yr. Conclusion is same as with P-value method.

Section 9.3

1. (i) (a) $\alpha = 0.01$

$H_0: p = 0.301$

$H_1: p < 0.301$

(b) Use the standard normal distribution. The $\hat{p}$ distribution is approximately normal when n is sufficiently large, which it is here, since $np = 215(0.301) \approx 647$ and $nq = 215(0.699) \approx 150.3$ are both > 5.

$$\hat{p} = \frac{r}{n} = \frac{46}{215} \approx 0.214$$

$$z = \frac{\hat{p} - p}{\sqrt{\frac{pq}{n}}} = \frac{0.214 - 0.301}{\sqrt{\frac{0.301(0.699)}{215}}} \approx -2.78$$

(c) P-value $= P(z < -2.78) = 0.0027$

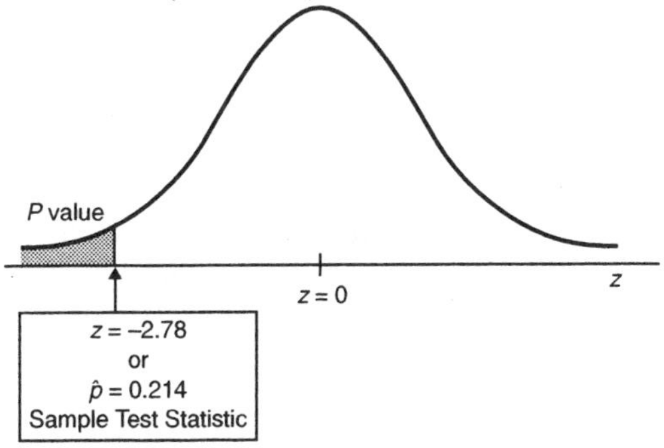

(d) Since a P-value of $0.0027 \le 0.01$ for α, we reject H_0. Yes.

(e) At the 1% level of significance, the sample data indicate the population proportion of numbers with a leading "1" in the revenue file is less than the 0.301 predicted by Benford's Law.

(ii) Yes; the revenue data file seems to include more numbers with higher first non-zero digits than Benford's Law predicts.

(iii) We have not proved H_0 to be false. However, because our sample data leads us to reject H_0, and conclude that there are too few numbers with leading digit "1," more investigation is merited.

3. (a) $\alpha = 0.01$
$H_0: p = 0.70$
$H_1: p \neq 0.70$

(b) Use the standard normal distribution. The $\hat{p}$ distribution is approximately normal when n is sufficiently large, which it is here, since $np = 32(0.7) = 22.4$ and $nq = 32(0.3) = 9.6$ are both > 5.

$$\hat{p} = \frac{r}{n} = \frac{24}{32} \approx 0.75$$

$$z = \frac{\hat{p} - p}{\sqrt{\frac{pq}{n}}} = \frac{0.75 - 0.70}{\sqrt{\frac{0.70(0.30)}{32}}} \approx 0.62$$

(c) P-value $= 2P(z > 0.62) = 2(0.2676) = 0.5352$

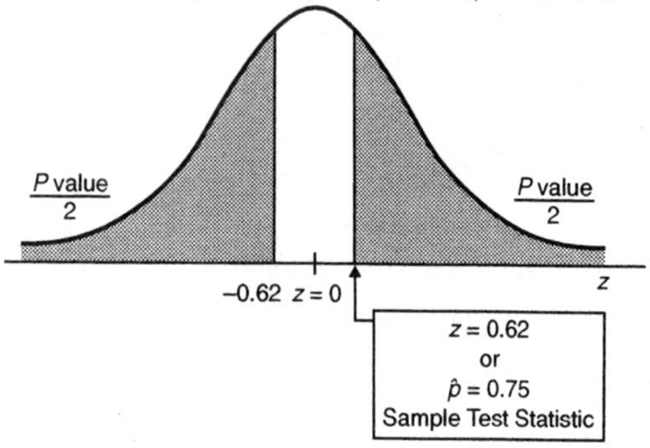

(d) Since a P-value of $0.5352 > 0.01$ for α, we fail to reject H_0. No.

(e) At the 1% level of significance, we cannot say that the population proportion of arrests of males aged 15 to 34 in Rock Springs is different from 70%.

5. (a) $\alpha = 0.01$
$H_0: p = 0.77$
$H_1: p < 0.77$

(b) Use the standard normal distribution. The $\hat{p}$ distribution is approximately normal when n is sufficiently large, which it is here, because $np = 27(0.77) = 20.79$ and $nq = 27(0.23) = 6.21$ are both > 5.

$$\hat{p} = \frac{r}{n} = \frac{15}{27} = 0.5556$$

$$z = \frac{\hat{p} - p}{\sqrt{\frac{pq}{n}}} = \frac{0.5556 - 0.77}{\sqrt{\frac{0.77(0.23)}{27}}} = -2.65$$

(c) P-value = $P(z > -2.65) = 0.0004$

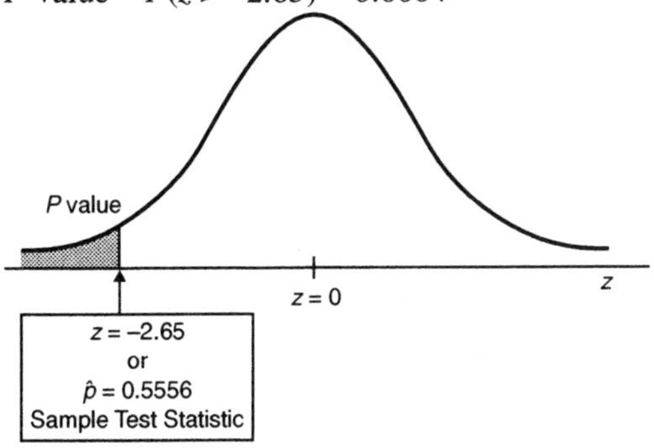

(d) Since a P-value of $0.0004 \le 0.01$ for α, we reject H_0. Yes.

(e) At the 1% level of significance, data show that the population proportion of driver fatalities related to alcohol is less than 77% in Kit Carson County.

7. (a) $\alpha = 0.01$
$H_0\colon p = 0.50$
$H_1\colon p < 0.50$

(b) Use the standard normal distribution. The $\hat{p}$ distribution is approximately normal when n is sufficiently large, and it is here, because $np = 34(0.50) = 17$ and $nq = 17$ are both > 5.

$$\hat{p} = \frac{r}{n} = \frac{10}{34} = 0.2941$$

$$z = \frac{\hat{p} - p}{\sqrt{\frac{pq}{n}}} = \frac{0.2941 - 0.50}{\sqrt{\frac{0.5(0.5)}{34}}} = -2.40$$

(c) P-value = $P(z < -2.40) = 0.0082$

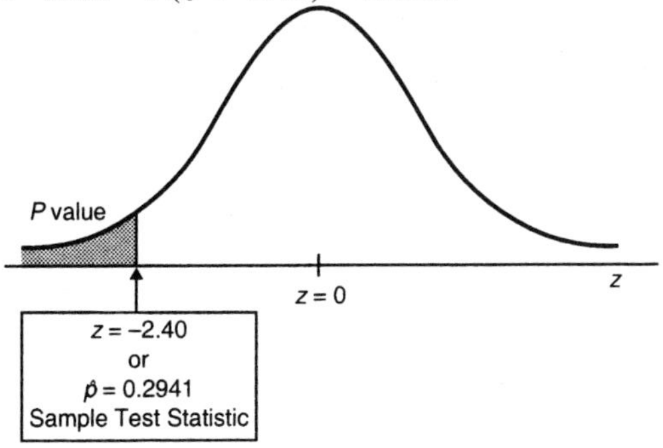

(d) Since a P-value of $0.0082 \le 0.01$ for α, we reject H_0. Yes.

(e) At the 1% level of significance, the data indicate that the population proportion of female wolves is now less than 50% in the region.

9. (a) $\alpha = 0.01$
$H_0\colon p = 0.261$
$H_1\colon p \ne 0.261$

(b) Use the standard normal distribution. The $\hat{p}$ distribution is approximately normal when n is sufficiently large, which it is here, because $np = 317(0.261) = 82.737$ and $nq = 317(0.739) = 234.263$ are both > 5.

$$\hat{p} = \frac{r}{n} = \frac{61}{317} = 0.1924$$

$$z = \frac{\hat{p} - p}{\sqrt{\frac{pq}{n}}} = \frac{0.1924 - 0.261}{\sqrt{\frac{0.261(0.739)}{317}}} \approx -2.78$$

(c) P-value $= 2P(z < -2.78) = 2(0.0027) = 0.0054$

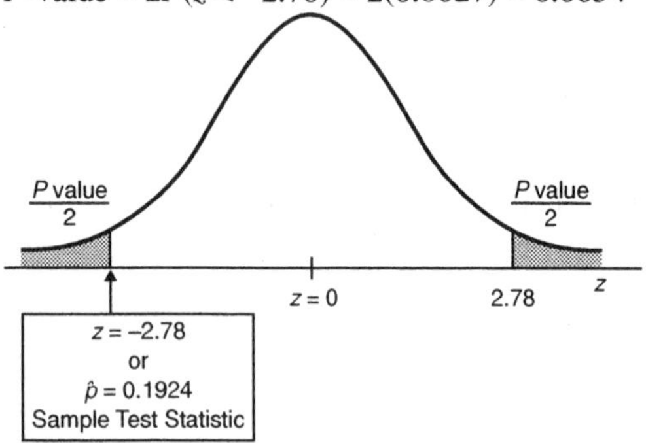

(d) Since a P-value of $0.0054 \le 0.01$ for α, we reject H_0. Yes.

(e) At the 1% level of significance, the sample data indicate that the population proportion of the five-syllable sequence is different from that of the text of Plato's *Republic*.

11. (a) $\alpha = 0.01$
H_0: $p = 0.47$
H_1: $p > 0.47$

(b) Use the standard normal distribution. The $\hat{p}$ distribution is approximately normal when n is sufficiently large, which it is here, because $np = 1006(0.47) = 472.82$ and $nq = 1006(0.53) = 533.18$ are both > 5.

$$\hat{p} = \frac{r}{n} = \frac{490}{1006} = 0.4871$$

$$z = \frac{\hat{p} - p}{\sqrt{\frac{pq}{n}}} = \frac{0.4871 - 0.47}{\sqrt{\frac{0.47(0.53)}{1006}}} = 1.09$$

(c) P-value $= P(z > 1.09) = 0.1379$

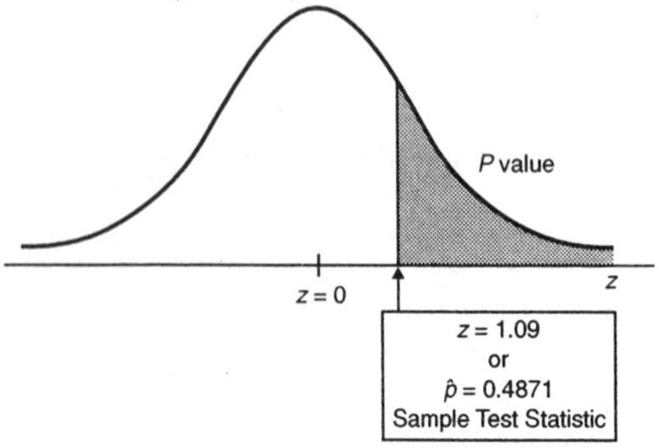

(d) Since a P-value of $0.1379 > 0.01$ for α, we fail to reject H_0. No.

(e) At the 1% level of significance, there is insufficient evidence to support the claim that the population proportion of customers loyal to Chevrolet is more than 47%.

13. (a) $\alpha = 0.05$
$H_0: p = 0.092$
$H_1: p > 0.092$

(b) Use the standard normal distribution. The $\hat{p}$ distribution is approximately normal when n is sufficiently large, which it is here, because $np = 196(0.092) = 18.032$ and $nq = 196(0.908) = 177.968$ are both > 5.

$$\hat{p} = \frac{r}{n} = \frac{29}{196} = 0.1480$$

$$z = \frac{\hat{p} - p}{\sqrt{\frac{pq}{n}}} = \frac{0.1480 - 0.092}{\sqrt{\frac{0.092(0.908)}{196}}} = 2.71$$

(c) P-value $= P(z > 2.71) = 0.0034$

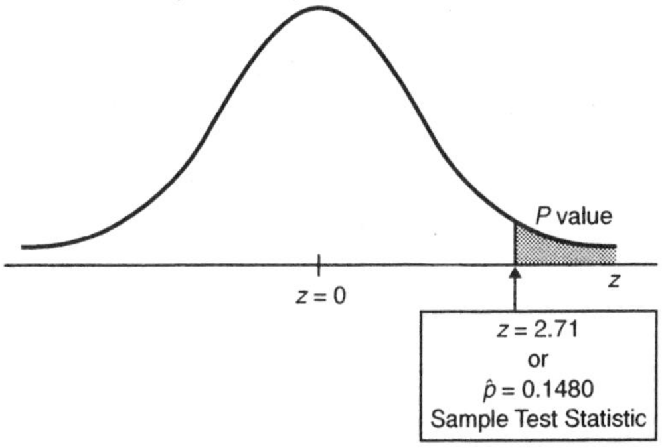

(d) Since a P-value of $0.0034 \le 0.05$ for α, we reject H_0. Yes.

(e) At the 5% level of significance, the data indicate that the population proportion of students with hypertension during final exams week is higher than 9.2%.

15. (a) $\alpha = 0.01$
$H_0: p = 0.82$
$H_1: p \neq 0.82$

(b) Use the standard normal distribution. The $\hat{p}$ distribution is approximately normal when n is sufficiently large, which it is here, because $np = 73(0.82) = 59.86$ and $nq = 73(0.18) = 13.14$ are both > 5.

$$\hat{p} = \frac{r}{n} = \frac{56}{73} = 0.7671$$

$$z = \frac{\hat{p} - p}{\sqrt{\frac{pq}{n}}} = \frac{0.7671 - 0.82}{\sqrt{\frac{0.82(0.18)}{73}}} = -1.18$$

(c) P-value $= 2P(z \le -1.18) = 2(0.1190) = 0.2380$

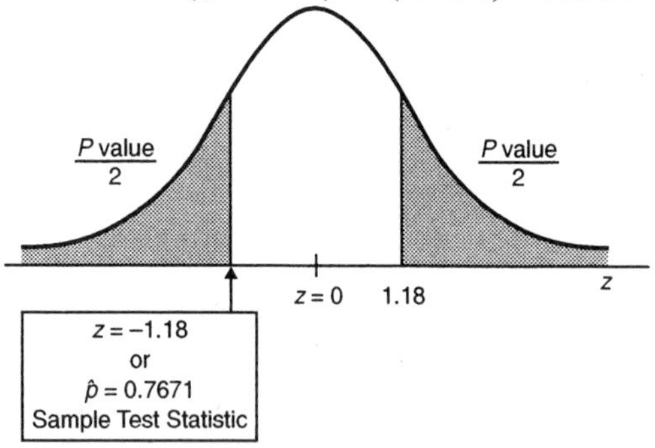

(d) Since a P-value of $0.2380 > 0.01$ for α, we fail to reject H_0. No.

(e) At the 1% level of significance, the evidence is insufficient to indicate that the population proportion of extroverts among college student government leaders is different from 82%.

17. (a) $\alpha = 0.01$
H_0: $p = 0.76$
H_1: $p \ne 0.76$

(b) Use the standard normal distribution. The $\hat{p}$ distribution is approximately normal when n is sufficiently large, which it is here, because $np = 59(0.76) = 44.84$ and $nq = 59(0.24) = 14.16$ are both > 5.

$$\hat{p} = \frac{r}{n} = \frac{47}{59} = 0.7966$$

$$z = \frac{\hat{p} - p}{\sqrt{\frac{pq}{n}}} = \frac{0.7966 - 0.76}{\sqrt{\frac{0.76(0.24)}{59}}} = 0.66$$

(c) P-value $= 2P(z > 0.66) = 2(0.2546) = 0.5092$

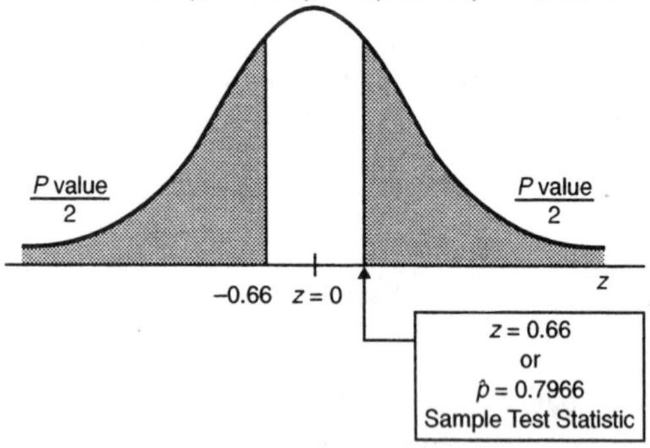

(d) Since a P-value of $0.5092 > 0.01$ for α, we fail to reject H_0. No.

(e) At the 1% level of significance, the evidence is insufficient to conclude that the population proportion of professors in Colorado who would choose the career again is different from the national rate of 76%.

19. For a left-tailed test and $\alpha = 0.01$, critical value is $z_0 = -2.33$. The critical region consists of values less than -2.33. The sample test statistic $z = -2.65$ is in the critical region so we reject H_0. Result is consistent with P-value conclusion.

Section 9.4

1. (a) $\alpha = 0.05$
$H_0: \mu_d = 0$
$H_1: \mu_d \neq 0$
Since $\neq$ is in H_1, a two-tailed test is used.

(b) Use the Student's t distribution. Assume d has a normal distribution or has a mound-shaped symmetric distribution.

Pair	1	2	3	4	5	6	7	8
$d = B - A$	3	−2	5	4	10	−15	6	7

$\bar{d} = 2.25,\ s_d = 7.78$

$$t = \frac{\bar{d} - \mu_d}{\frac{s_d}{\sqrt{n}}} = \frac{2.25 - 0}{\frac{7.78}{\sqrt{8}}} = 0.818$$

(c) $d.f. = n - 1 = 8 - 1 = 7$
From Table 6, 0.818 falls between entries 0.711 and 1.254. Using two-tailed areas, find that $0.250 < P\text{-value} < 0.500$.

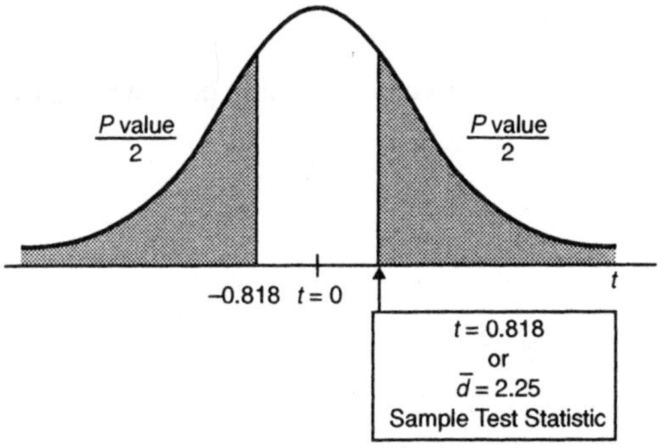

(d) Since the P-value interval is > 0.05 for α, we fail to reject H_0. No.

(e) At the 5% level of significance, the evidence is insufficient to claim a difference in population mean percentage increases for corporate revenue and CEO salary.

3. (a) $\alpha = 0.01$
$H_0: \mu_d = 0$
$H_1: \mu_d > 0$
Since > is in H_1, a right-tailed test is used.

(b) Use the Student's t distribution. Assume d has a normal distribution or has a mound-shaped symmetric distribution.

Pair	1	2	3	4	5
d = Jan – April	35	9	26	–24	17

$\bar{d} = 12.6,\ s_d = 22.66$

$$t = \frac{\bar{d} - \mu_d}{\frac{s_d}{\sqrt{n}}} = \frac{12.6 - 0}{\frac{22.66}{\sqrt{5}}} = 1.243$$

(c) $d.f. = n - 1 = 5 - 1 = 4$
From Table 6, 1.243 falls between entries 0.741 and 1.344. Using one-tailed areas, find that $0.125 < P\text{-value} < 0.250$.

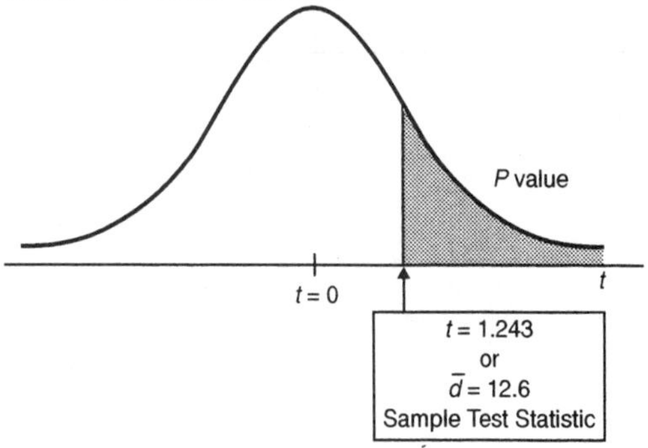

(d) Since the P-value interval is > 0.01 for α, we fail to reject H_0. No.

(e) At the 1% level of significance, the evidence is insufficient to claim average peak wind gusts are higher in January.

5. (a) $\alpha = 0.05$
$H_0\colon \mu_d = 0$
$H_1\colon \mu_d > 0$
Since > is in H_1, a right-tailed test is used.

(b) Use the Student's t distribution. Assume d has a normal distribution or has a mound-shaped symmetric distribution.

Pair	1	2	3	4	5	6	7	8
d = winter – summer	19	–4	17	7	9	0	–9	10

$\bar{d} = 6.125,\ s_d = 9.83$

$$t = \frac{\bar{d} - \mu_d}{\frac{s_d}{\sqrt{n}}} = \frac{6.125 - 0}{\frac{9.83}{\sqrt{8}}} = 1.76$$

(c) $d.f. = n - 1 = 8 - 1 = 7$
From Table 6, 1.76 falls between entries 1.617 and 1.895. Using one-tailed areas, find that $0.05 < P\text{-value} < 0.075$.

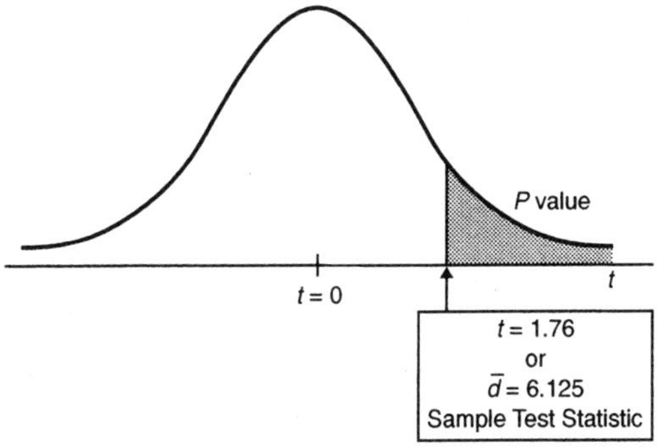

(d) Since the P-value interval is > 0.05 for α, we fail to reject H_0. No.

(e) At the 5% level of significance, the evidence is insufficient to indicate the population average percentage of male wolves in winter is higher.

7. (a) $\alpha = 0.05$
H_0: $\mu_d = 0$
H_1: $\mu_d > 0$
Since $>$ is in H_1, a right-tailed test is used.

(b) Use the Student's t distribution. Assume d has a normal distribution or has a mound-shaped symmetric distribution.

Pair	1	2	3	4	5	6	7	8
d = houses – hogans	5	2	22	–23	–4	–19	33	32

$\bar{d} = 6$, $s_d = 21.5$

$$t = \frac{\bar{d} - \mu_d}{\frac{s_d}{\sqrt{n}}} = \frac{6 - 0}{\frac{21.5}{\sqrt{8}}} = 0.789$$

(c) $d.f. = n - 1 = 8 - 1 = 7$
From Table 6, 0.789 falls between entries 0.711 and 1.254. Use one-tailed areas to find that $0.125 < P\text{-value} < 0.250$.

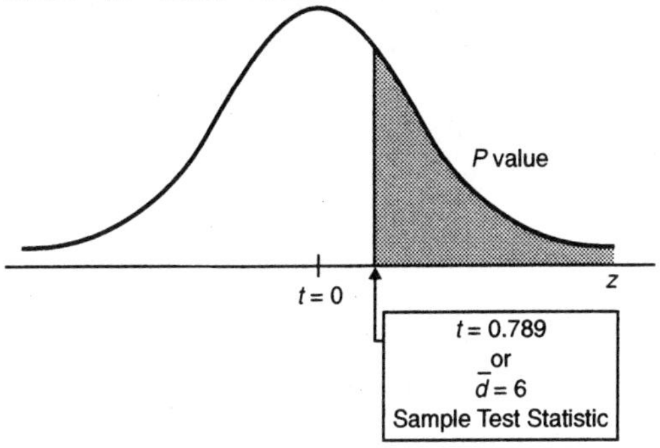

(d) Since the P-value interval is > 0.05 for α, we reject H_0. No.

(e) At the 5% level of significance, the evidence is insufficient to show the mean number of inhabited houses is greater than that of hogans.

9. (a) $\alpha = 0.05$
H_0: $\mu_d = 0$
H_1: $\mu_d \neq 0$
Since ≠ is in H_1, a two-tailed test is used.

(b) Use the Student's t distribution. Assume d has a normal distribution or has a mound-shaped symmetric distribution.

Pair	1	2	3	4	5
$d = 1 - 2$	6	3	−6	−3	5

$\bar{d} = 1.0,\ s_d = 5.24$

$$t = \frac{\bar{d} - \mu_d}{\frac{s_d}{\sqrt{n}}} = \frac{1.0 - 0}{\frac{5.24}{\sqrt{5}}} = 0.427$$

(c) $d.f. = n - 1 = 5 - 1 = 4$
From Table 6, 0.427 falls to the left of (is smaller than) any entry in $d.f.$ = 4 row. Use two-tailed areas to find that P-value > 0.500.

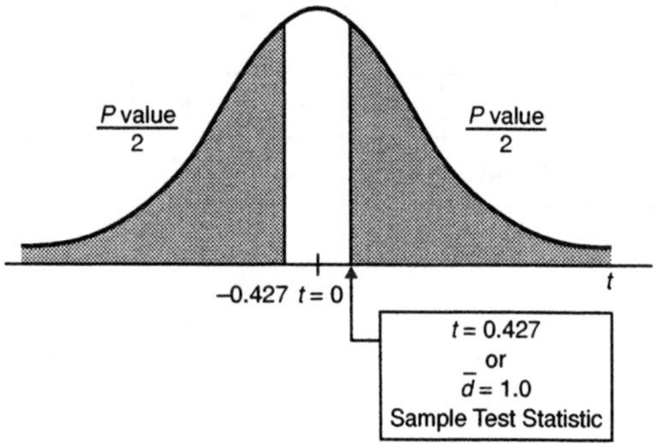

(d) Since the P-value is > 0.05 for α, we fail to reject H_0. No.

(e) At the 5% level of significance, the evidence is insufficient to indicate a difference in the population mean number of service ware sherds in subarea 1 as compared with subarea 2.

11. (i) Use a calculator to verify. Rounded values are used in part ii.

(ii) (a) $\alpha = 0.05$
H_0: $\mu = 0$
H_1: $\mu > 0$
Since > is in H_1, a right-tailed test is used.

(b) Use the Student's t distribution. The sample size is greater than 30.

$\bar{d} = 2.472,\ s_d = 12.124$

$$t = \frac{\bar{d} - \mu_d}{\frac{s_d}{\sqrt{n}}} = \frac{2.472 - 0}{\frac{12.124}{\sqrt{36}}} \approx 1.223$$

(c) $d.f. = n - 1 = 36 - 1 = 35$

From Table 6, 1.223 falls between entries 1.170 and 1.306. Use one-tailed areas to find that $0.100 < P\text{-value} < 0.125$.

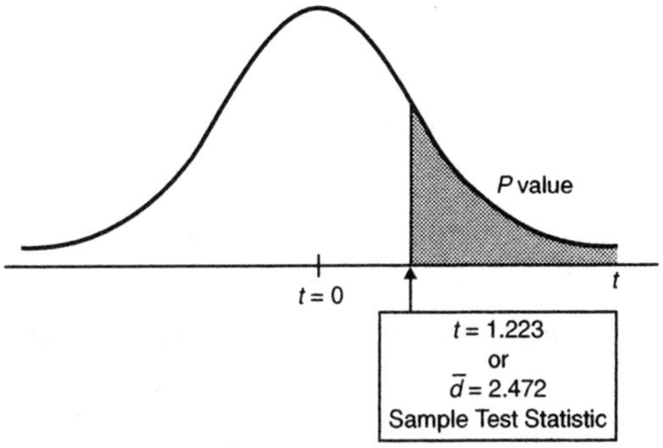

(d) Since the P-value interval > 0.05 for α, we fail to reject H_0. No.

(e) At the 5% level of significance, the evidence is insufficient to claim that the population mean cost of living index for housing is higher than that for groceries.

13. (a) $\alpha = 0.05$

$H_0\colon \mu_d = 0$

$H_1\colon \mu_d > 0$

Since > is in H_1, a right-tailed test is used.

(b) Use the Student's t distribution. Assume d has a normal distribution or has a mound-shaped symmetric distribution.

Pair	1	2	3	4	5	6	7	8	9
$d = B - A$	7	−2	9	0	6	1	−3	−3	3

$\bar{d} = 2.0,\ s_d = 4.5$

$$t = \frac{\bar{d} - \mu_d}{\frac{s_d}{\sqrt{n}}} = \frac{2.0 - 0}{\frac{4.5}{\sqrt{9}}} = 1.33$$

(c) $d.f. = n - 1 = 9 - 1 = 8$
From Table 6, 1.33 falls between entries 1.240 and 1.397. Use one-tailed areas to find that $0.100 < P\text{-value} < 0.125$.

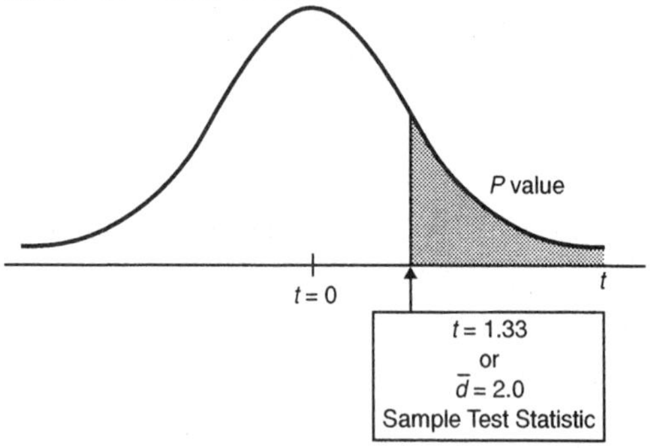

(d) Since the P-value interval is > 0.05 for α, we fail to reject H_0. No.

(e) At the 5% level of significance, the evidence is insufficient to claim that the population score on the last round is higher than that on the first.

15. (a) $\alpha = 0.05$
H_0: $\mu_d = 0$
H_1: $\mu_d > 0$
Since > is in H_1, a right-tailed test is used.

(b) Use the Student's t distribution. Assume d has a normal distribution or has a mound-shaped symmetric distribution.

Pair	1	2	3	4	5	6	7	8
d = time 1 – time 5	1.4	1.7	–0.8	1.5	–0.5	–0.1	1.7	1.3

$\bar{d} = 0.775,\ s_d = 1.0539$

$$t = \frac{\bar{d} - \mu_d}{\frac{s_d}{\sqrt{n}}} = \frac{0.775 - 0}{\frac{1.0539}{\sqrt{8}}} = 2.080$$

(c) $d.f. = n - 1 = 8 - 1 = 7$
From Table 6, 2.080 falls between entries 1.895 and 2.365. Use one-tailed areas to find $0.025 < P\text{-value} < 0.050$.

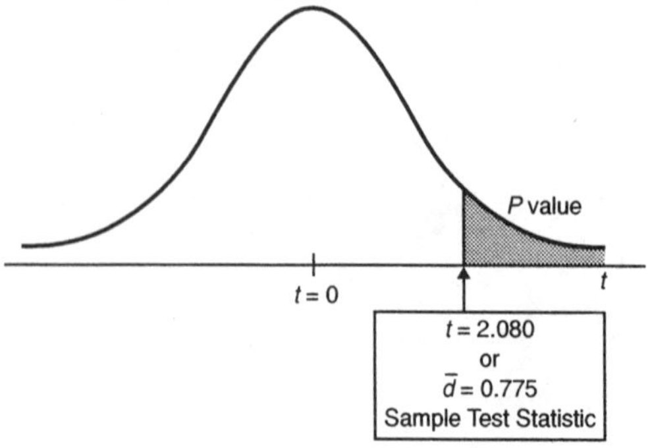

(d) Since the P-value interval is ≤ 0.05 for α, we reject H_0. Yes.

(e) At the 5% level of significance, the evidence is sufficient to claim that the population mean time for rats receiving larger rewards to climb the ladder is less.

17. For a two-tailed test with $\alpha = 0.05$ and $d.f. = 7$, critical values are $\pm t_0 = \pm 2.365$. The sample test statistic $t = 0.818$ falls between -2.365 and 2.365, so we do not reject H_0. This conclusion is the same as that reached by the P-value method.

Section 9.5

1. (a) $\alpha = 0.01$
$H_0\colon \mu_1 = \mu_2$
$H_1\colon \mu_1 > \mu_2$

(b) Use the standard normal distribution. We assume both population distributions are approximately normal and σ_1 and σ_2 are known.
$\bar{x}_1 - \bar{x}_2 = 2.8 - 2.1 = 0.7$

$$z = \frac{(\bar{x}_1 - \bar{x}_2) - (\mu_1 - \mu_2)}{\sqrt{\frac{\sigma_1^2}{n_1} + \frac{\sigma_2^2}{n_2}}} = \frac{0.7 - 0}{\sqrt{\frac{0.5^2}{10} + \frac{0.7^2}{10}}} \approx 2.57$$

(c) P-value $= P(z > 2.57) = 0.0051$

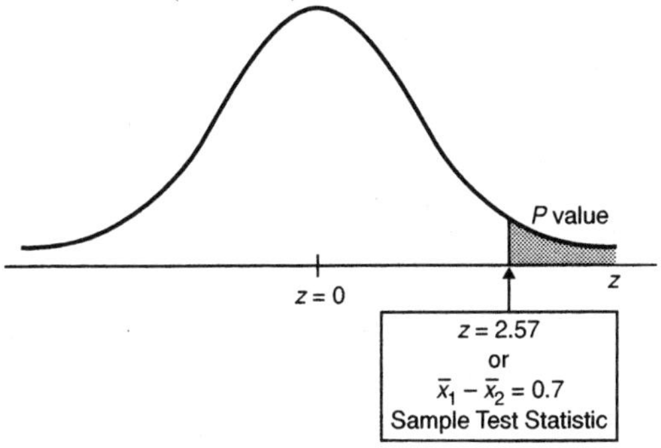

(d) Since P-value $= 0.0051 \leq 0.01$ for α, we reject H_0. Yes.

(e) At the 1% level of significance, the evidence is sufficient to indicate that the population mean REM sleep time for children is more than for adults.

3. (a) $\alpha = 0.05$
$H_0\colon \mu_1 = \mu_2$
$H_1\colon \mu_1 \neq \mu_2$

(b) Use the standard normal distribution. We assume both population distributions are approximately normal and σ_1 and σ_2 are known.

$\bar{x}_1 - \bar{x}_2 = 4.9 - 4.3 = 0.6$

$$z = \frac{(\bar{x}_1 - \bar{x}_2) - (\mu_1 - \mu_2)}{\sqrt{\frac{\sigma_1^2}{n_1} + \frac{\sigma_2^2}{n_2}}} = \frac{0.6 - 0}{\sqrt{\frac{1.5^2}{46} + \frac{1.2^2}{51}}} \approx 2.16$$

(c) P-value $= 2P(z > 2.16) = 2(0.0154) = 0.0308$

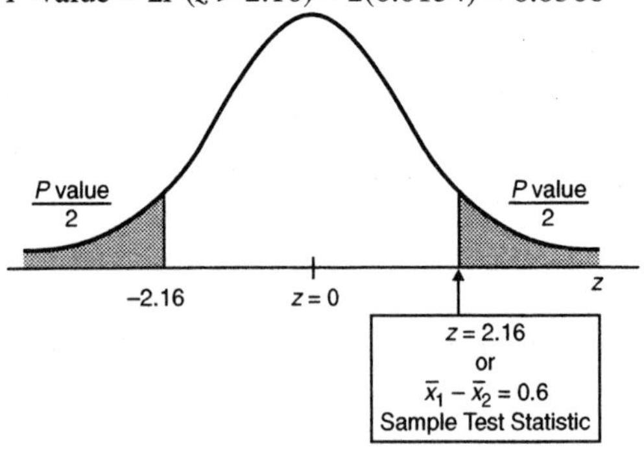

(d) Since P-value $= 0.0308 \le 0.05$ for α, we reject H_0. Yes.

(e) At the 5% level of significance, the evidence is sufficient to show there is a difference between mean response regarding preference for camping or preference for fishing.

5. (i) Use a calculator to verify. Use rounded results to compute t.

(ii) (a) $\alpha = 0.01$

H_0: $\mu_1 = \mu_2$

H_1: $\mu_1 < \mu_2$

(b) Use the Student's t distribution. We assume both population distributions are approximately normal and σ_1 and σ_2 are unknown.

$\bar{x}_1 - \bar{x}_2 = 3.51 - 3.87 = -0.36$

$$t = \frac{(\bar{x}_1 - \bar{x}_2) - (\mu_1 - \mu_2)}{\sqrt{\frac{s_1^2}{n_1} + \frac{s_2^2}{n_2}}} = \frac{-0.36 - 0}{\sqrt{\frac{0.81^2}{10} + \frac{0.94^2}{12}}} \approx -0.965$$

(c) Since $n_1 < n_2$, $d.f. = n_1 - 1 = 10 - 1 = 9$
In Table 6, 0.965 falls between entries 0.703 and 1.230. Use one-tailed areas to find that $0.125 < P\text{-value} < 0.250$.

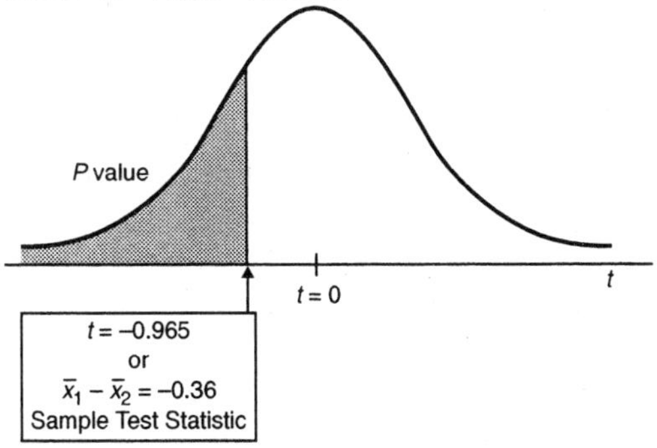

(d) Since the P-value interval > 0.01 for α, we fail to reject H_0. No.

(e) At the 1% level of significance, the evidence is insufficient to indicate that violent crime in the Rocky Mountain region is higher than in New England.

7. (a) $\alpha = 0.05$
$H_0\colon \mu_1 = \mu_2$
$H_1\colon \mu_1 \neq \mu_2$

(b) Use the Student's t distribution. Both sample sizes are large, $n_i \geq 30$, and σ_1 and σ_2 are unknown.
$\bar{x}_1 - \bar{x}_2 = 344.5 - 354.2 = -9.7$

$$t = \frac{(\bar{x}_1 - \bar{x}_2) - (\mu_1 - \mu_2)}{\sqrt{\frac{s_1^2}{n_1} + \frac{s_2^2}{n_2}}} = \frac{-9.7 - 0}{\sqrt{\frac{49.1^2}{30} + \frac{50.9^2}{30}}} \approx -0.751$$

(c) Since $n_1 = n_2$, $d.f. = n - 1 = 30 - 1 = 29$.
From Table 6, 0.751 falls between entries 0.683 and 1.174. Use two-tailed areas to find $0.250 < P\text{-value} < 0.500$.

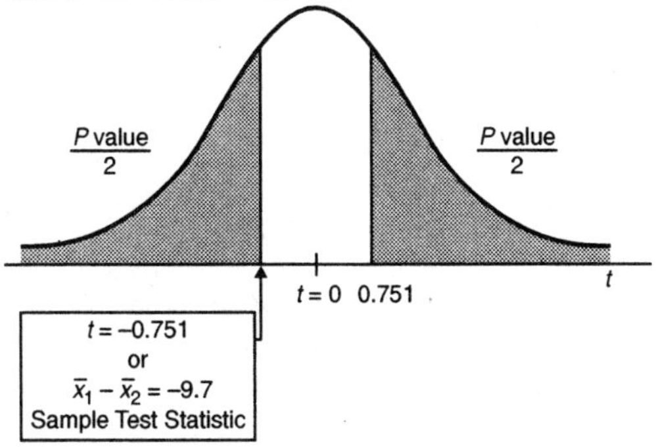

(d) Since the P-value interval > 0.05 for α, we fail to reject H_0. No.

(e) At the 5% level of significance, the evidence is insufficient to indicate there is a difference between the control and experimental group in the mean score on the vocabulary portion of the test.

9. (i) Use a calculator to verify. Use rounded results to compute t.

(ii) (a) $\alpha = 0.05$
$H_0: \mu_1 = \mu_2$
$H_1: \mu_1 \neq \mu_2$

(b) Use the Student's t distribution. We assume the population distributions for both are approximately normal and mound shaped, and σ_1 and σ_2 are unknown.

$\bar{x}_1 - \bar{x}_2 = 4.75 - 3.93 = 0.82$

$$t = \frac{(\bar{x}_1 - \bar{x}_2) - (\mu_1 - \mu_2)}{\sqrt{\frac{s_1^2}{n_1} + \frac{s_2^2}{n_2}}} = \frac{0.82 - 0}{\sqrt{\frac{2.82^2}{16} + \frac{2.43^2}{15}}} \approx 0.869$$

(c) Since $n_2 < n_1$, $d.f. = n_2 - 1 = 15 - 1 = 14$.
From Table 6, 0.869 falls between entries 0.692 and 1.200. Use two-tailed areas to find $0.250 < P\text{-value} < 0.500$.

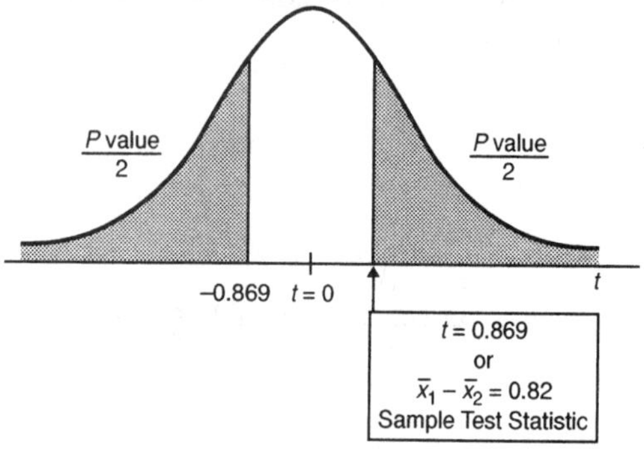

(d) Since the P-value interval is > 0.05 for α, we fail to reject H_0. No.

(e) At the 5% level of significance, the evidence is insufficient to indicate that there is a difference in the mean number of cases of fox rabies between the two regions.

11. (i) Use a calculator to verify. Use rounded results to compute t.

(ii) (a) $\alpha = 0.05$
$H_0: \mu_1 = \mu_2$
$H_1: \mu_1 \neq \mu_2$

(b) Use the Student's t distribution. We assume the population distributions for both are approximately normal and mound shaped, and σ_1 and σ_2 are unknown.

$\bar{x}_1 - \bar{x}_2 = 4.86 - 6.5 = -1.64$

$$t = \frac{(\bar{x}_1 - \bar{x}_2) - (\mu_1 - \mu_2)}{\sqrt{\frac{s_1^2}{n_1} + \frac{s_2^2}{n_2}}} = \frac{-1.64 - 0}{\sqrt{\frac{3.18^2}{7} + \frac{2.88^2}{8}}} \approx -1.041$$

(c) Since $n_1 < n_2$, $d.f. = n_1 - 1 = 7 - 1 = 6$.

From Table 6, 1.041 falls between entries 0.718 and 1.273. Use two-tailed areas to find $0.250 < P\text{-value} < 0.500$.

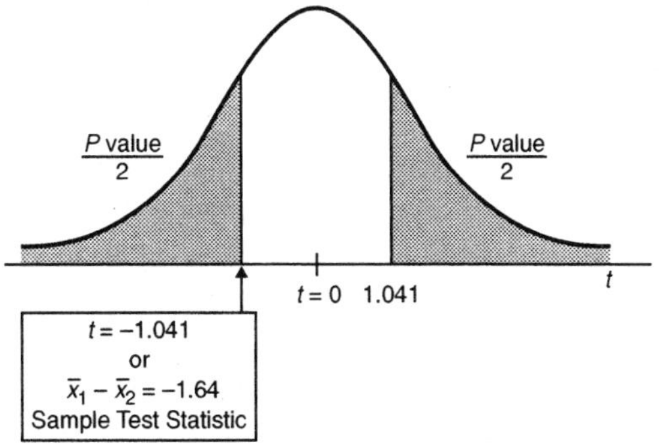

(d) Since the P-value interval is > 0.05 for α, we fail to reject H_0. No.

(e) At the 5% level of significance, the evidence is insufficient to indicate that the mean time lost due to hot tempers is different from that lost due to technical worker's attitudes.

13. **(i)** Use a calculator to verify. Use rounded results to compute t.

(ii) **(a)** $\alpha = 0.01$

H_0: $\mu_1 = \mu_2$

H_1: $\mu_1 < \mu_2$

(b) Use the Student's t distribution. We assume the population distributions for both are approximately normal and mound shaped, and σ_1 and σ_2 are unknown.

$\bar{x}_1 - \bar{x}_2 = 7.2 - 10.8 = -3.6$

$$t = \frac{(\bar{x}_1 - \bar{x}_2) - (\mu_1 - \mu_2)}{\sqrt{\frac{s_1^2}{n_1} + \frac{s_2^2}{n_2}}} = \frac{-3.6 - 0}{\sqrt{\frac{2.5^2}{11} + \frac{2.5^2}{12}}} \approx -3.450$$

(c) Since $n_1 < n_2$, $d.f. = 11 - 1 = 10$.

From Table 6, 3.450 falls between entries 3.169 and 4.587. Use one-tailed areas to find $0.0005 < P\text{-value} < 0.005$.

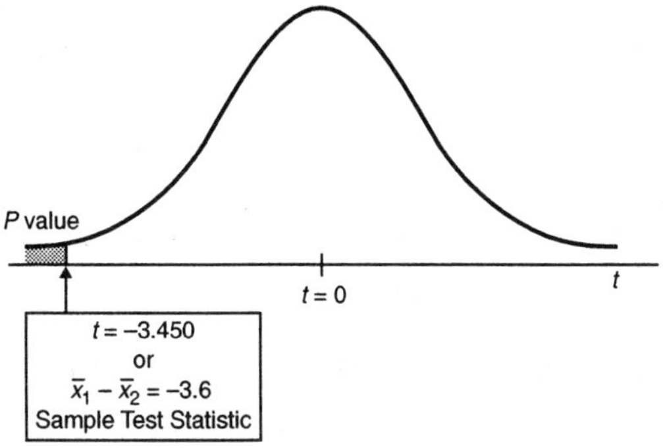

(d) Since the P-value interval is ≤ 0.01 for α, we reject H_0. Yes.

(e) At the 1% level of significance, the evidence is sufficient to indicate that the mean increase in water temperature at the surface has increased since the addition of the new generator.

15. (a) $$d.f. \approx \frac{\left(\frac{s_1^2}{n_1}+\frac{s_2^2}{n_2}\right)^2}{\frac{1}{n_1-1}\left(\frac{s_1^2}{n_1}\right)^2+\frac{1}{n_2-1}\left(\frac{s_2^2}{n_2}\right)^2} = \frac{\left(\frac{0.81^2}{10}+\frac{0.94^2}{12}\right)^2}{\frac{1}{9}\left(\frac{0.81^2}{10}\right)^2+\frac{1}{11}\left(\frac{0.94^2}{12}\right)^2} \approx 19.96$$

(Note: Some software will truncate this to 19.)

(b) In Problem 5, $d.f. = n_1 - 1 = 10 - 1 = 9$.

The convention of using the smaller of $n_1 - 1$ and $n_2 - 1$ leads to a *d.f.* that is always less than or equal to that computed by Satterthwaite's formula.

17. (a) $\alpha = 0.05$
$H_0\colon p_1 = p_2$
$H_1\colon p_1 \neq p_2$

(b) Use the standard normal distribution. The number of trials is sufficiently large because $n_1\bar{p}$, $n_1\bar{q}$, $n_2\bar{p}$, and $n_2\bar{q}$ are each larger than 5.

$$\bar{p} = \frac{r_1 + r_2}{n_1 + n_2} = \frac{59+56}{220+175} \approx 0.2911$$
$$\bar{q} = 1 - \bar{p} = 1 - 0.2911 = 0.7089$$
$$\hat{p}_1 = \frac{r_1}{n_1} = \frac{59}{220} \approx 0.268,\ \hat{p}_2 = \frac{r_2}{n_2} = \frac{56}{175} = 0.32$$
$$\hat{p}_1 - \hat{p}_2 = 0.268 - 0.32 = -0.052$$
$$z = \frac{\hat{p}_1 - \hat{p}_2}{\sqrt{\frac{\bar{p}\bar{q}}{n_1}+\frac{\bar{p}\bar{q}}{n_2}}} = \frac{-0.052}{\sqrt{\frac{0.2911(0.7089)}{220}+\frac{0.2911(0.7089)}{175}}} \approx -1.13$$

(c) P-value $= 2P(z < -1.13) = 2(0.1292) = 0.2584$

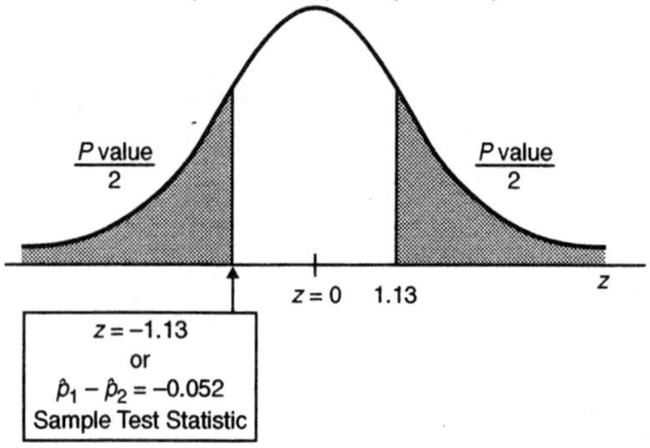

(d) Since the P-value $= 0.2584 > 0.05$ for α, we fail to reject H_0. No.

(e) At the 1% level of significance, there is insufficient evidence to conclude that the population proportion of women favoring more tax dollars for the arts is different from the proportion of men.

19. (a) $\alpha = 0.01$
$H_0: p_1 = p_2$
$H_1: p_1 \neq p_2$

(b) Use the standard normal distribution. The number of trials is sufficiently large because $n_1\overline{p}$, $n_1\overline{q}$, $n_2\overline{p}$, and $n_2\overline{q}$ are each larger than 5.

$$\overline{p} = \frac{r_1 + r_2}{n_1 + n_2} = \frac{12+7}{153+128} = 0.0676$$

$$\overline{q} = 1 - \overline{p} = 1 - 0.0676 = 0.9324$$

$$\hat{p}_1 = \frac{r_1}{n_1} = \frac{12}{153} \approx 0.0784,\ \hat{p}_2 = \frac{r_2}{n_2} = \frac{7}{128} \approx 0.0547$$

$$\hat{p}_1 - \hat{p}_2 = 0.0784 - 0.0547 = 0.0237$$

$$z = \frac{\hat{p}_1 - \hat{p}_2}{\sqrt{\frac{\overline{p}\overline{q}}{n_1} + \frac{\overline{p}\overline{q}}{n_2}}} = \frac{0.0237}{\sqrt{\frac{0.0676(0.9324)}{153} + \frac{0.0676(0.9324)}{128}}} \approx 0.79$$

(c) P-value $= 2P(z > 0.79) = 2(0.2148) = 0.4296$

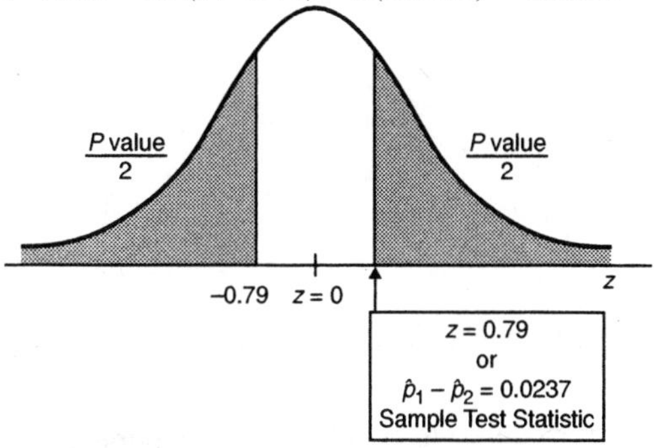

(d) Since the P-value $= 0.4296 > 0.01$ for α, we fail to reject H_0. No.

(e) At the 1% level of significance, there is insufficient evidence to conclude that the population proportion of high school dropouts on Oahu is different from that of Sweetwater County.

21. (a) $\alpha = 0.01$
$H_0: p_1 = p_2$
$H_1: p_1 < p_2$

(b) Use the standard normal distribution. The number of trials is sufficiently large because $n_1\bar{p}$, $n_1\bar{q}$, $n_2\bar{p}$, and $n_2\bar{q}$ are each larger than 5.

$$\bar{p} = \frac{r_1 + r_2}{n_1 + n_2} = \frac{37+47}{100+100} = 0.42$$

$$\bar{q} = 1 - \bar{p} = 1 - 0.42 = 0.58$$

$$\hat{p}_1 = \frac{r_1}{n_1} = \frac{37}{100} = 0.37,\ \hat{p}_2 = \frac{r_2}{n_2} = \frac{47}{100} = 0.47$$

$$\hat{p}_1 - \hat{p}_2 = 0.37 - 0.47 = -0.10$$

$$z = \frac{\hat{p}_1 - \hat{p}_2}{\sqrt{\frac{\bar{p}\bar{q}}{n_1} + \frac{\bar{p}\bar{q}}{n_2}}} = \frac{-0.10}{\sqrt{\frac{0.42(0.58)}{100} + \frac{0.42(0.58)}{100}}} \approx -1.43$$

(c) P-value $= P(z < -1.43) \approx 0.0764$

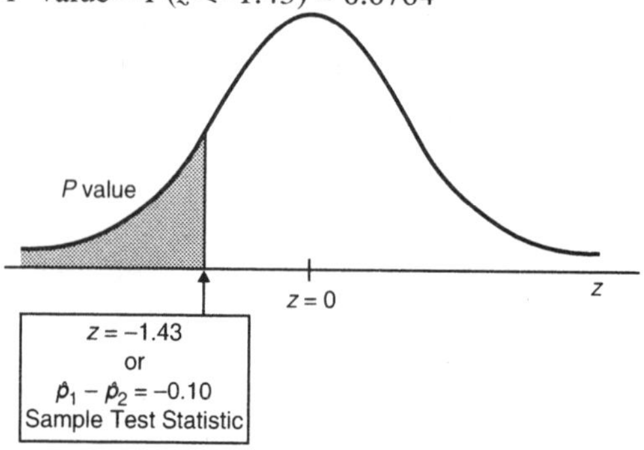

(d) Since the P-value $= 0.0764 > 0.01$ for α, we fail to reject H_0. No.

(e) At the 1% level of significance, there is insufficient evidence to conclude that the population proportion of adults believing in extraterrestrials who attended college is higher than the proportion who did not attend college.

23. (a) $\alpha = 0.01$

H_0: $p_1 = p_2$

H_1: $p_1 < p_2$

(b) Use the standard normal distribution. The number of trials is sufficiently large because $n_1\bar{p}$, $n_1\bar{q}$, $n_2\bar{p}$, and $n_2\bar{q}$ are each larger than 5.

$$\bar{p} = \frac{r_1 + r_2}{n_1 + n_2} = \frac{194+320}{378+516} \approx 0.5749$$

$$\bar{q} = 1 - \bar{p} = 1 - 0.5749 = 0.4251$$

$$\hat{p}_1 = \frac{r_1}{n_1} = \frac{194}{378} \approx 0.5132,\ \hat{p}_2 = \frac{r_2}{n_2} = \frac{320}{516} \approx 0.6202$$

$$\hat{p}_1 - \hat{p}_2 = 0.5132 - 0.6202 = -0.1070$$

$$z = \frac{\hat{p}_1 - \hat{p}_2}{\sqrt{\frac{\bar{p}\bar{q}}{n_1} + \frac{\bar{p}\bar{q}}{n_2}}} = \frac{-0.1070}{\sqrt{\frac{0.5749(0.4251)}{378} + \frac{0.5749(0.4251)}{516}}} \approx -3.19$$

(c) P-value $= 2P(z < -3.19) = 0.0007$

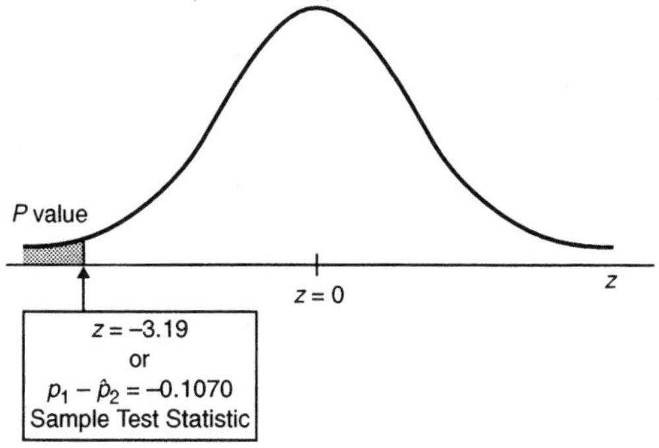

(d) Since the P-value $= 0.0007 \le 0.01$ for α, we reject H_0. Yes.

(e) At the 1% level of significance, there is sufficient evidence to conclude that the population proportion of guests requesting nonsmoking rooms has increased.

25. (a) $\alpha = 0.05$
$H_0: p_1 = p_2$
$H_1: p_1 < p_2$

(b) Use the standard normal distribution. The number of trials is sufficiently large because $n_1\bar{p}$, $n_1\bar{q}$, $n_2\bar{p}$, and $n_2\bar{q}$ are each larger than 5.

$$\bar{p} = \frac{r_1 + r_2}{n_1 + n_2} = \frac{45 + 71}{250 + 280} \approx 0.2189$$

$$\bar{q} = 1 - \bar{p} = 1 - 0.2189 = 0.7811$$

$$\hat{p}_1 = \frac{r_1}{n_1} = \frac{45}{250} = 0.18,\ \hat{p}_2 = \frac{r_2}{n_2} = \frac{71}{280} \approx 0.2536$$

$$\hat{p}_1 - \hat{p}_2 = 0.18 - 0.2536 \approx -0.074$$

$$z = \frac{\hat{p}_1 - \hat{p}_2}{\sqrt{\frac{\bar{p}\bar{q}}{n_1} + \frac{\bar{p}\bar{q}}{n_2}}} = \frac{-0.074}{\sqrt{\frac{0.2189(0.7811)}{250} + \frac{0.2189(0.7811)}{280}}} \approx -2.04$$

(c) P-value $= P(z < -2.04) = 0.0207$

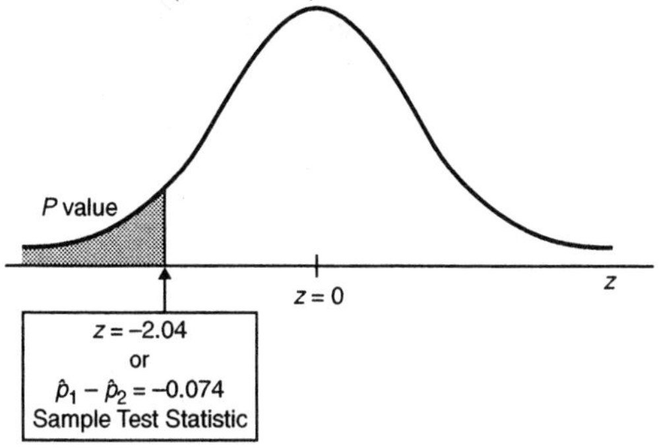

(d) Since the P-value $= 0.0207 \le 0.05$ for α, we reject H_0. Yes.

(e) At the 5% level of significance, there is sufficient evidence to conclude that the population proportion of trusting people in Chicago is higher in the older group.

Chapter 9 Review

1. (a) $\alpha = 0.05$
H_0: $\mu = 11.1$
H_1: $\mu \neq 11.1$
Since $\neq$ is in H_1, use a two-tailed test.

(b) Use the standard normal distribution. We assume x has a normal distribution with known standard deviation σ. Note that μ is given in the null hypothesis.

$$z = \frac{\bar{x} - \mu}{\frac{\sigma}{\sqrt{n}}} = \frac{10.8 - 11.1}{\frac{0.6}{\sqrt{36}}} = -3.00$$

(Note: 600 miles = 0.6 thousand miles)

(c) P-value $= 2P(z < -3.00) = 2(0.0013) = 0.0026$

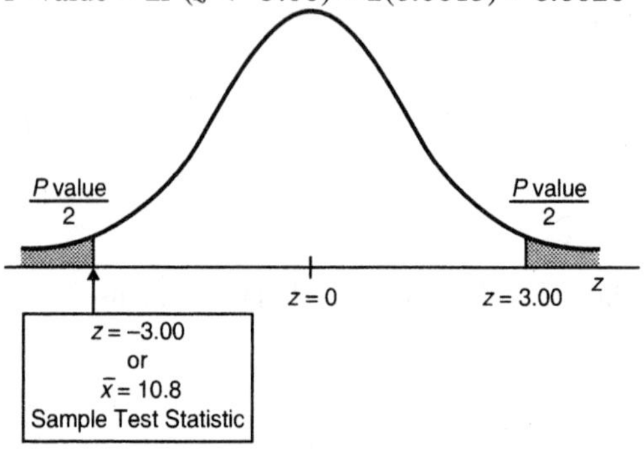

(d) Since a P-value of $0.0026 \leq 0.05$ for α, we reject H_0. Yes.

(e) At the 5% level of significance, the evidence is sufficient to say that the miles driven per vehicle in Chicago is different than the national average.

3. (a) $\alpha = 0.01$
H_0: $\mu = 0.8$ A
H_1: $\mu > 0.8$ A

(b) Use the Student's t distribution with $d.f. = n - 1 = 9 - 1 = 8$. We assume x has a distribution that is approximately normal and σ is unknown.

$$t = \frac{\bar{x} - \mu}{\frac{s}{\sqrt{n}}} = \frac{1.4 - 0.8}{\frac{0.41}{\sqrt{9}}} \approx 4.390$$

(c) For $d.f. = 8$, 4.390 falls between entries 3.355 and 5.041. Using one-tailed areas, $0.0005 < P\text{-value} < 0.005$.

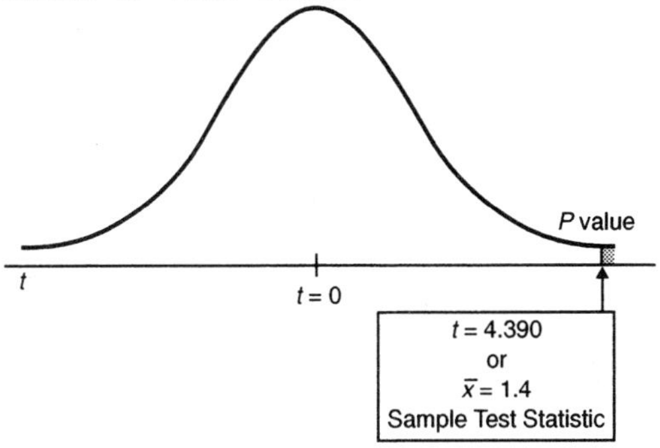

(d) Since the entire P-value interval ≤ 0.01 for α, we reject H_0. Yes.

(e) At the 5% level of significance, the evidence is sufficient to say the Toylot claim of 0.8 A is too low.

5. (a) $\alpha = 0.01$
$H_0: p = 0.60$
$H_1: p < 0.60$

(b) Use the standard normal distribution. The $\hat{p}$ distribution is approximately normal when n is sufficiently large, which it is here, because $np = 90(0.6) = 54$ and $nq = 90(0.4) = 36$ are both > 5.

$$\hat{p} = \frac{r}{n} = \frac{40}{90} = 0.4444$$

$$z = \frac{\hat{p} - p}{\sqrt{\frac{pq}{n}}} = \frac{0.4444 - 0.60}{\sqrt{\frac{0.60(0.40)}{90}}} = -3.01$$

(c) $P\text{-value} = P(z < -3.01) = 0.0013$

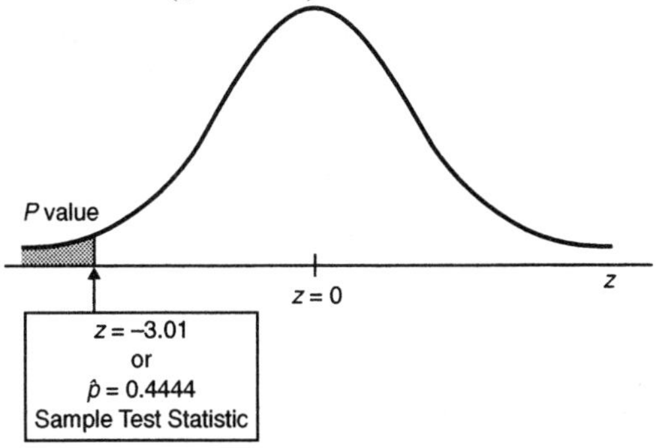

(d) Since a P-value of $0.0013 \le 0.01$ for α, we reject H_0. Yes.

(e) At the 1% level of significance, the evidence is sufficient to show the mortality rate has dropped.

7. (a) $\alpha = 0.01$
$H_0: p = 0.20$
$H_1: p > 0.20$

(b) Use the standard normal distribution. The $\hat{p}$ distribution is approximately normal when n is sufficiently large, which it is here because $np = 200(0.2) = 40$ and $nq = 200(0.8) = 160$ are both > 5.

$$\hat{p} = \frac{r}{n} = \frac{56}{200} = 0.28$$

$$z = \frac{\hat{p} - p}{\sqrt{\frac{pq}{n}}} = \frac{0.28 - 0.20}{\sqrt{\frac{0.20(0.80)}{200}}} = 2.83$$

(c) P-value $= P(z > 2.83) = 0.0023$

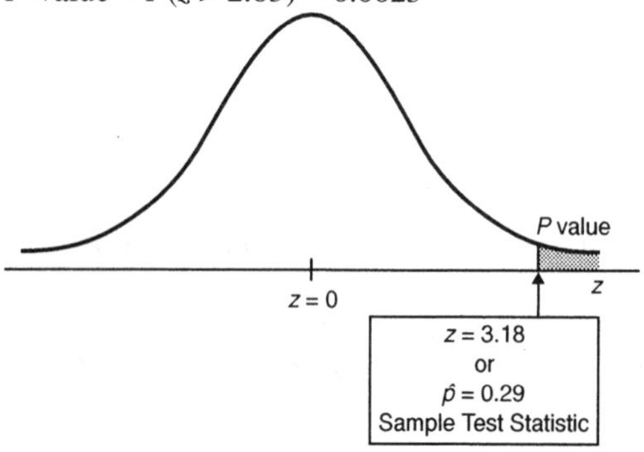

(d) Since a P-value of $0.0023 \le 0.01$ for α, we reject H_0. Yes.

(e) At the 1% level of significance the evidence is sufficient to show the population proportion of students who read the magazine is larger than 0.20.

9. (a) $\alpha = 0.01$
H_0: $\mu = 40$
H_1: $\mu > 40$
Since $>$ is in H_1, use a right-tailed test.

(b) Use the standard normal distribution. We assume x has a normal distribution with known standard deviation σ. Note that μ is given in the null hypothesis.

$$z = \frac{\bar{x} - \mu}{\frac{\sigma}{\sqrt{n}}} = \frac{43.1 - 40}{\frac{9}{\sqrt{94}}} \approx 3.34$$

(c) P-value $= P(z > 3.34) = 0.0004$

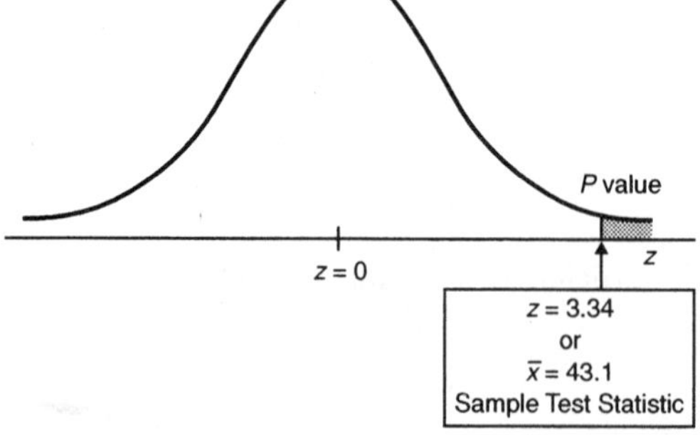

(d) Since a P-value of $0.0004 \le 0.01$ for α, we reject H_0. Yes.

(e) At the 1% level of significance, the evidence is sufficient to say the population average number of matches is larger than 40.

11. (a) $\alpha = 0.05$
H_0: $\mu_1 = \mu_2$
H_1: $\mu_1 \neq \mu_2$

(b) Use the Student's *t* distribution. We assume both population distributions are approximately normal and σ_1 and σ_2 are unknown.
$\bar{x}_1 - \bar{x}_2 = 4.8 - 5.8 = -1$

$$t = \frac{(\bar{x}_1 - \bar{x}_2) - (\mu_1 - \mu_2)}{\sqrt{\frac{s_1^2}{n_1} + \frac{s_2^2}{n_2}}} = \frac{-1}{\sqrt{\frac{2^2}{16} + \frac{1.8^2}{14}}} \approx -1.441$$

(c) Since $n_2 < n_1$, $d.f. = n_2 - 1 = 14 - 1 = 13$.
In Table 6, 1.441 falls between entries 1.350 and 1.530. Use two-tailed areas to find that $0.150 < P\text{-value} < 0.200$.

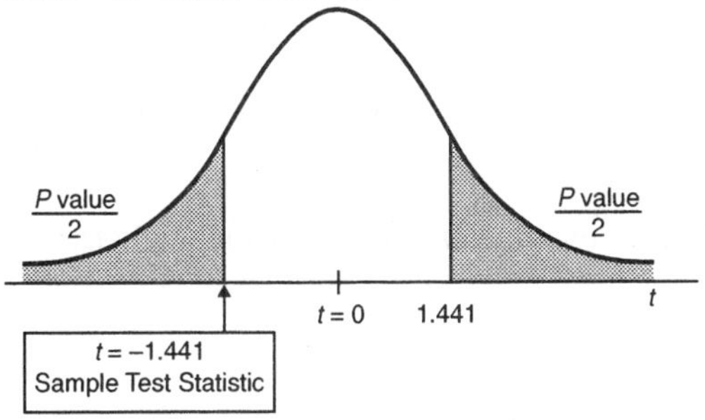

(d) Since the *P*-value interval is > 0.05 for α, we fail to reject H_0. No.

(e) At the 1% level of significance, the evidence is insufficient to show that there is a difference in the population mean waiting times.

13. (a) $\alpha = 0.05$
H_0: $\mu = 7$ oz
H_1: $\mu \neq 7$ oz

(b) Use the Student's *t* distribution with $d.f. = n - 1 = 8 - 1 = 7$. We assume *x* has a distribution that is approximately normal and σ is unknown.

$$t = \frac{\bar{x} - \mu}{\frac{s}{\sqrt{n}}} = \frac{7.3 - 7}{\frac{0.5}{\sqrt{8}}} \approx 1.697$$

(c) For $d.f. = 7$, 1.697 falls between entries 1.617 and 1.895. Using two-tailed areas, $0.100 < P\text{-value} < 0.150$.

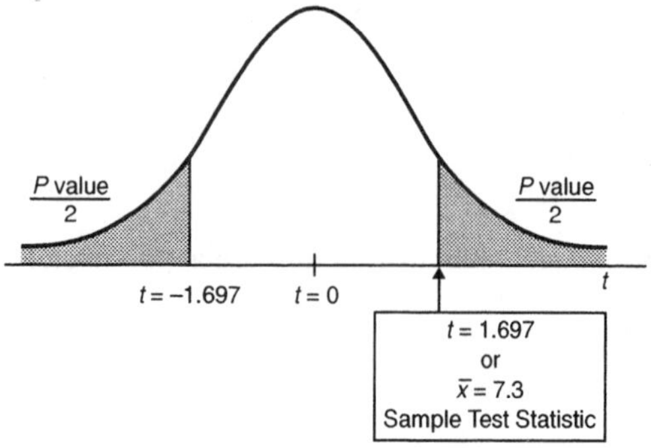

(d) Since the entire P-value interval > 0.05 for α, we fail to reject H_0. No.

(e) At the 5% level of significance, the evidence is insufficient to show that the population mean amount of coffee per cup is different from 7 ounces.

15. (a) $\alpha = 0.05$
$H_0: \mu_d = 0$
$H_1: \mu_d < 0$

(b) Use the Student's t distribution. Assume d has a normal distribution or has a mound-shaped symmetric distribution.

Pair	1	2	3	4	5
d = Before − After	−6.4	−7.2	−8.6	−3.8	1.3

$\bar{d} = -4.94,\ s_d = 3.90$

$$t = \frac{\bar{d} - \mu_d}{\frac{s_d}{\sqrt{n}}} = \frac{-4.94 - 0}{\frac{3.90}{\sqrt{5}}} = -2.832$$

(c) $d.f. = n - 1 = 5 - 1 = 4$
From Table 6, 2.832 falls between entries 2.776 and 3.747. Use one-tailed areas to find that $0.010 < P\text{-value} < 0.025$.

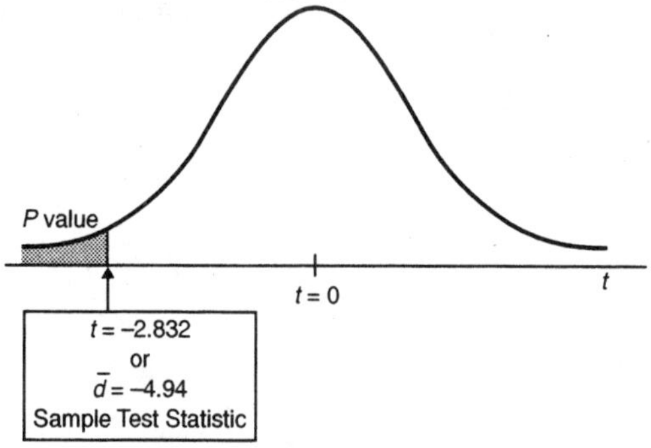

(d) Since the P-value interval is ≤ 0.05 for α, we reject H_0. Yes.

(e) At the 5% level of significance, there is insufficient evidence to claim that the injection system lasts less than an average of 48 months.

17. (a) $\alpha = 0.05$

H_0: $\mu_1 = \mu_2$

H_1: $\mu_1 \neq \mu_2$

(b) Use the Student's t distribution. We assume both population distributions are approximately normal and σ_1 and σ_2 are unknown.

$\bar{x}_1 - \bar{x}_2 = 3.0 - 2.7 = 0.3$

$$t = \frac{(\bar{x}_1 - \bar{x}_2) - (\mu_1 - \mu_2)}{\sqrt{\frac{s_1^2}{n_1} + \frac{s_2^2}{n_2}}} = \frac{0.3 - 0}{\sqrt{\frac{0.8^2}{55} + \frac{0.9^2}{51}}} \approx 1.808$$

(c) Since $n_2 < n_1$, $d.f. = n_2 - 1 = 51 - 1 = 50$.

In Table 6, 1.808 falls between entries 1.676 and 2.009. Use two-tailed areas to find that $0.050 < P\text{-value} < 0.100$.

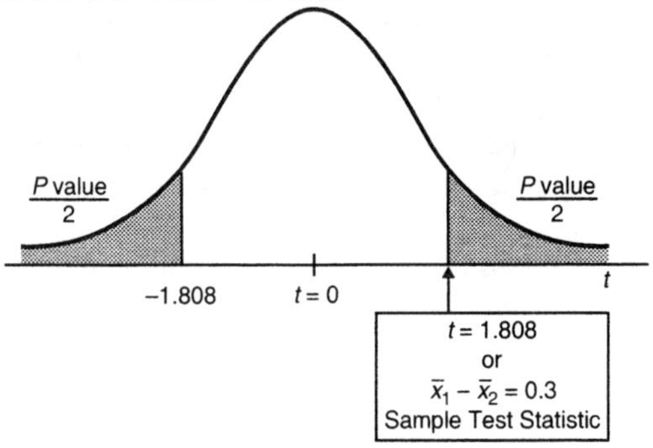

(d) Since the P-value interval > 0.05 for α, we fail to reject H_0. No.

(e) At the 5% level of significance, the evidence is insufficient to show that there is a difference in population mean length of the two types of projectile points.

Chapter 10 Correlation and Regression

Section 10.1

1. (a) The points seem close to a straight line, so there is moderate linear correlation.

(b) No straight line is realistically a good fit, so there is no linear correlation.

(c) The points seem very close to a straight line, so there is high linear correlation.

3. (a) No. The correlation coefficient is a mathematical tool for measuring the strength of a linear relationship between two variables. As such, it makes no implication about cause or effect. Just because two variables tend to increase or decrease together does not mean a change in one is *causing* a change in the other. A strong correlation between x and y is sometimes due to other (either known or unknown) variables. Such variables are called *lurking variables.*

(b) Increasing population might be a lurking variable causing both variables to increase.

5. (a) No. The correlation coefficient is a mathematical tool for measuring the strength of a linear relationship between two variables. As such, it makes no implication about cause or effect. Just because two variables tend to increase or decrease together does not mean a change in one is *causing* a change in the other. A strong correlation between x and y is sometimes due to other (either known or unknown) variables. Such variables are called *lurking variables.*

(b) One lurking variable for average annual income increases is inflation. Better training might be a lurking variable responsible for short times to run the mile.

7. (a)

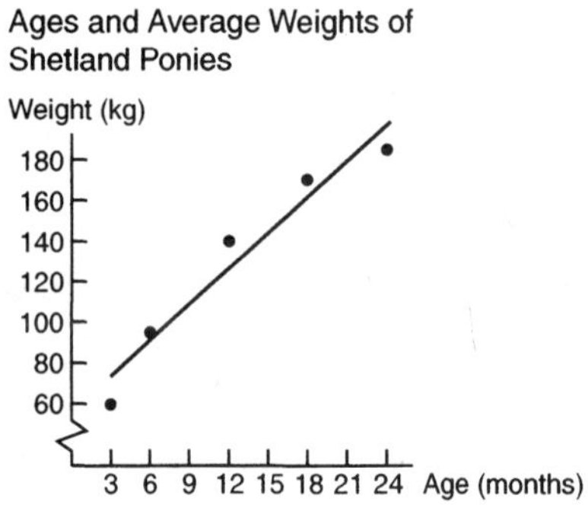

Draw the line you think fits best. (Method to find equation is in Section 10.2.)

(b) Since the points are very close to a straight line, the correlation is strong and positive (since it goes up from left to right).

(c) $$r = \frac{n\sum xy - (\sum x)(\sum y)}{\sqrt{n\sum x^2 - (\sum x)^2}\sqrt{n\sum y^2 - (\sum y)^2}} = \frac{5(9930) - (63)(650)}{\sqrt{5(1089) - (63)^2}\sqrt{5(95{,}350) - (650)^2}} \approx 0.972$$

Since r is positive, y should increase as x increases.

9. (a)

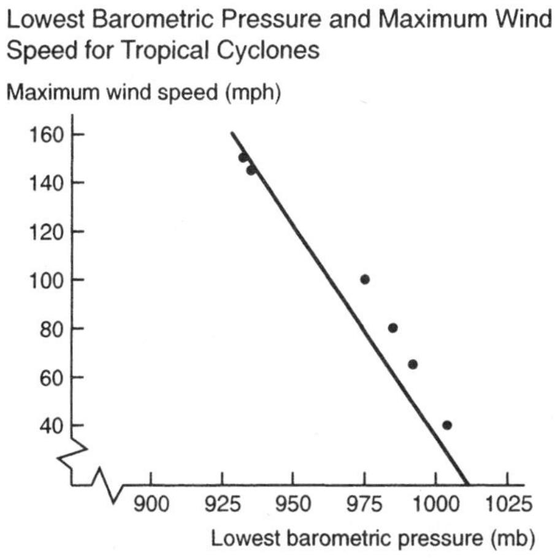

Draw the line you think fits best. (Method to find equation is in Section 10.2.)

(b) Since the points are very close to a straight line, the correlation is strong and negative (since it goes down from left to right).

(c) $$r = \frac{n\sum xy - (\sum x)(\sum y)}{\sqrt{n\sum x^2 - (\sum x)^2}\sqrt{n\sum y^2 - (\sum y)^2}} = \frac{6(556,315) - (5823)(580)}{\sqrt{6(5,655,779) - (5823)^2}\sqrt{6(65,750) - (580)^2}} \approx -0.990$$

Since r is negative, y should decrease as x increases.

11. (a)

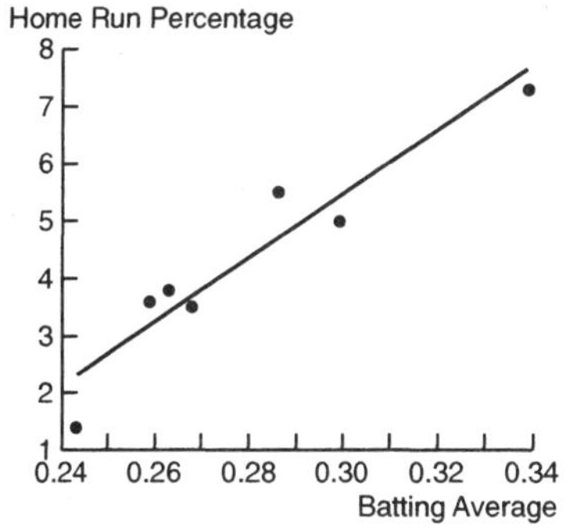

Draw the line you think fits best. (Method to find is in Section 10.2.)

(b) Since the points are very close to a straight line, the correlation is strong and positive (since it goes up from left to right).

(c) $$r = \frac{n\sum xy - (\sum x)(\sum y)}{\sqrt{n\sum x^2 - (\sum x)^2}\sqrt{n\sum y^2 - (\sum y)^2}} = \frac{7(8.753) - (1.957)(30.1)}{\sqrt{7(0.553) - (1.957)^2}\sqrt{7(150.15) - (30.1)^2}} \approx 0.948$$

Since r is positive, y should increase as x increases.

13. (a)

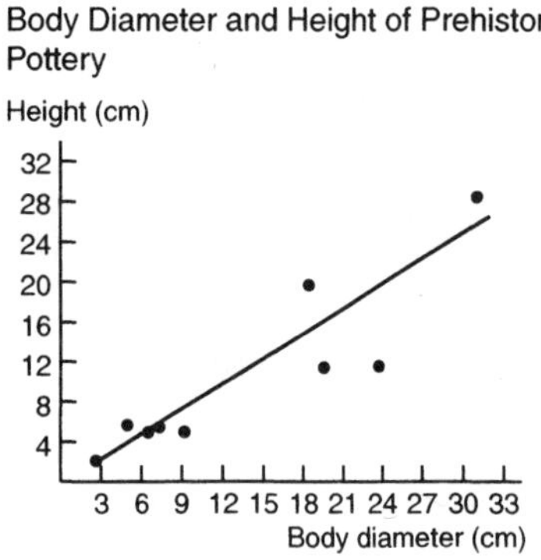

Draw the line you think fits best. (Method to find equation is in Section 10.2.)

(b) Since the points are fairly close to a straight line, the correlation is moderate and positive (since it goes up from left to right).

(c) $$r = \frac{n\sum xy - (\sum x)(\sum y)}{\sqrt{n\sum x^2 - (\sum x)^2}\sqrt{n\sum y^2 - (\sum y)^2}} = \frac{9(1897.19) - (123.1)(94.6)}{\sqrt{9(2452.45) - (123.1)^2}\sqrt{9(1584.3) - (94.6)^2}} \approx 0.896$$

Since r is positive, y should increase as x increases.

15. (a)

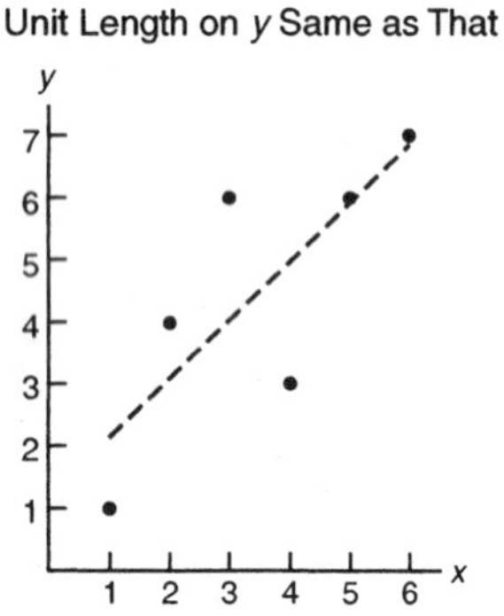

(b)

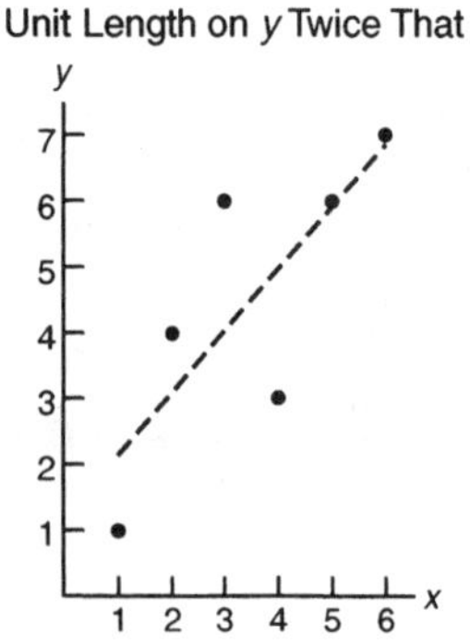

(c) 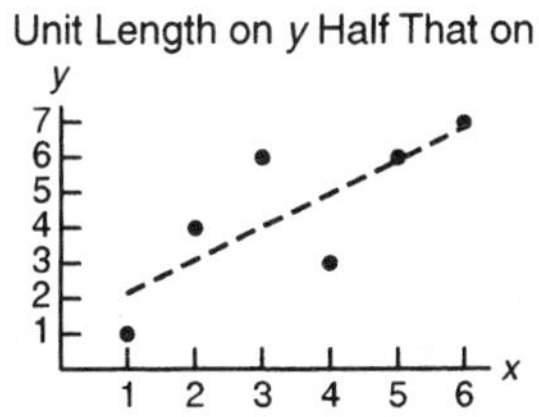

(d) Draw the lines you think best fit the data points.
Stretching the scale on the y-axis makes the line appear steeper. Shrinking the scale on the y-axis makes the line appear flatter. Or, shrinking the scale on the y-axis makes the line appear flatter. The slope of the line does not change. Only the appearance (visual impression) of slope changes as the scale of the y-axis changes.

17. (a) $r \approx 0.972$ with $n = 5$
Since $|0.972| = 0.972 \geq 0.88$ for $\alpha = 0.05$, r is significant for this α. For this α, we conclude that age and weight of Shetland ponies are correlated.

(b) $r \approx -0.990$ with $n = 6$
Since $|-0.990| = 0.990 \geq 0.92$ for $\alpha = 0.01$, r is significant for $\alpha = 0.01$.
For this α, we conclude that the lowest barometric pressure reading and maximum wind speed for cyclones are correlated.

19. (a) 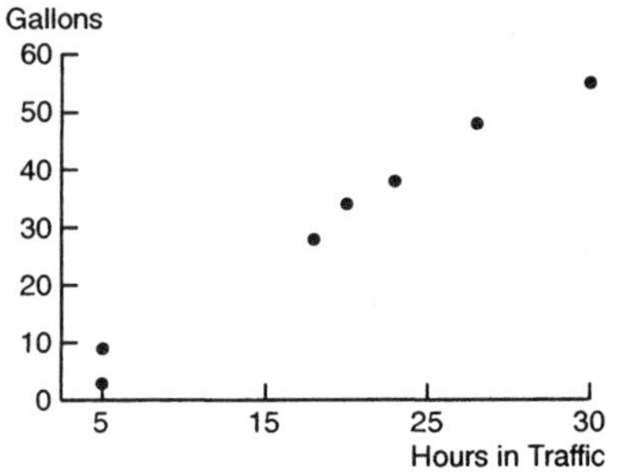

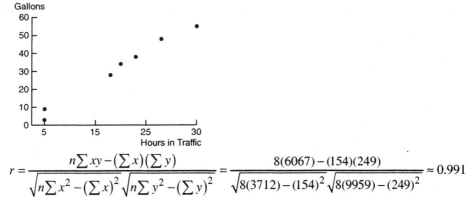

$$r = \frac{n\sum xy - (\sum x)(\sum y)}{\sqrt{n\sum x^2 - (\sum x)^2}\sqrt{n\sum y^2 - (\sum y)^2}} = \frac{8(6067) - (154)(249)}{\sqrt{8(3712) - (154)^2}\sqrt{8(9959) - (249)^2}} \approx 0.991$$

(b) For variables based on averages, $\overline{x} = 19.25$ hr; $s_x \approx 10.33$ hr; $\overline{y} = 31.13$ gal; $s_y \approx 17.76$ gal. For variables based on single individuals, $\overline{x} = 20.13$ hr; $s_x \approx 13.84$ hr; $\overline{y} = 31.87$ gal; $s_y \approx 25.18$; dividing by larger numbers results in a smaller value.

(c) 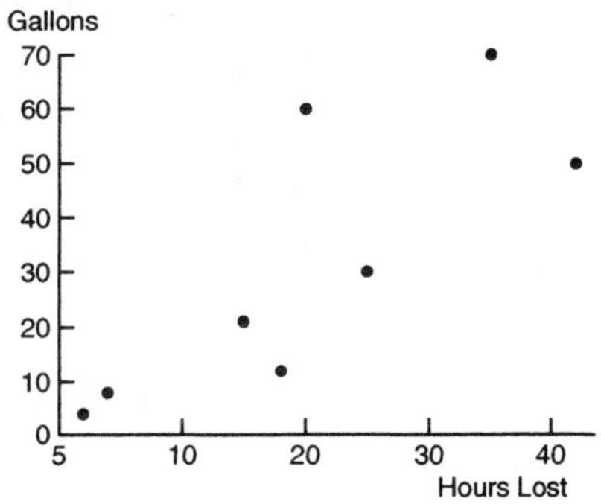

$$r = \frac{8(7071) - (161)(255)}{\sqrt{8(4583) - (161)^2}\sqrt{8(12{,}565) - (255)^2}} \approx 0.794$$

(d) $0.991 > 0.791$
Yes; by the central limit theorem, the $\overline{x}$ distribution has a smaller standard deviation than the corresponding x distribution.

Section 10.2

Note: In this section and the next two, answers may vary slightly, depending on how many significant digits are used throughout the calculations.

1. (a) Number of Jobs (in hundreds)

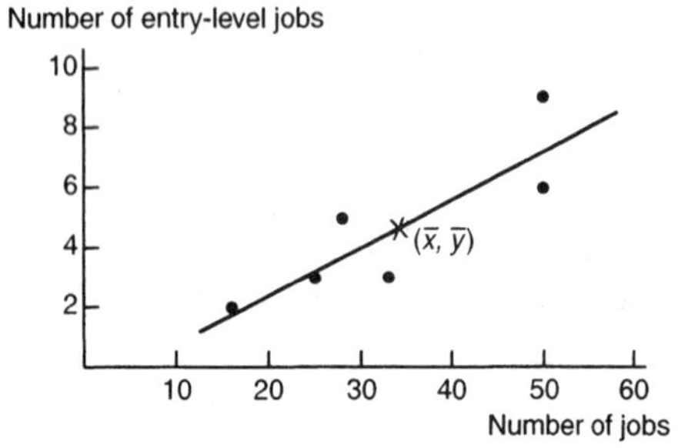

(b) Use a calculator to verify.

(c) $\overline{x} = \frac{\sum x}{n} = \frac{202}{6} \approx 33.67$ jobs

$\overline{y} = \frac{\sum y}{n} = \frac{28}{6} \approx 4.67$ entry-level jobs

$$b = \frac{n\sum xy - (\sum x)(\sum y)}{n\sum x^2 - (\sum x)^2} = \frac{6(1096) - (202)(28)}{6(7754) - (202)^2} \approx 0.161$$

$a = \overline{y} - b\overline{x} \approx 4.67 - (0.161)(33.67) \approx -0.748$

$\hat{y} = a + bx$ or $\hat{y} \approx -0.748 + 0.161x$

(d) See figure of part (a).

(e) $r^2 = (0.860)^2 \approx 0.740$
This means that 74.0% of the variation in y = number of entry-level jobs can be explained by the corresponding variation in x = total number of jobs using the least squares line.
100% − 74.0% = 26.0% of the variation is unexplained.

(f) Use $x = 40$.
$\hat{y} = -0.748 + 0.161(40)$
$\hat{y} = 5.69$ entry-level jobs

3. (a)

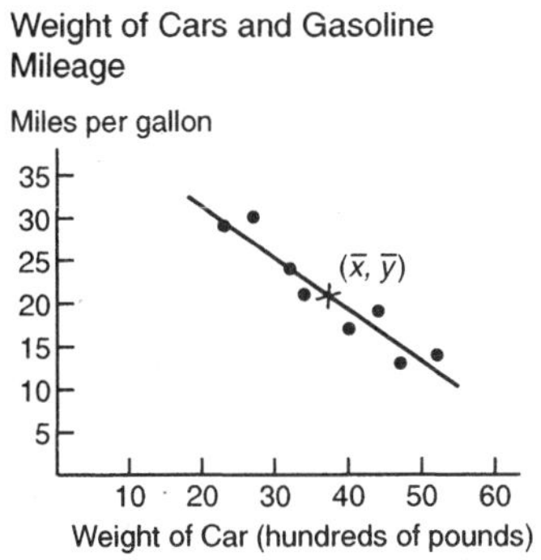

(b) Use a calculator to verify.

(c) $\bar{x} = \frac{\sum x}{n} = \frac{299}{8} = 37.375$

$\bar{y} = \frac{\sum y}{n} = \frac{167}{8} = 20.875$

$b = \frac{n\sum xy - (\sum x)(\sum y)}{n\sum x^2 - (\sum x)^2} = \frac{8(5814) - (299)(167)}{8(11,887) - (167)^2} \approx -0.6007$

$a = \bar{y} - b\bar{x} = 20.875 - (-0.6007)(37.375) \approx 43.326$

$\hat{y} = a + bx$ or $\hat{y} = 43.326 - 0.6007x$

(d) See figure of part (a).

(e) $r^2 = (-0.946)^2 \approx 0.895$

This means that 89.5% of the variation in y = miles per gallon can be explained by the corresponding variation in x = weight using the least squares line. 100% – 89.5% = 10.5% of the variation is unexplained.

(f) Use $x = 38$.

$\hat{y} = 43.326 - 0.6007(38) = 20.5$ mpg

5. (a)

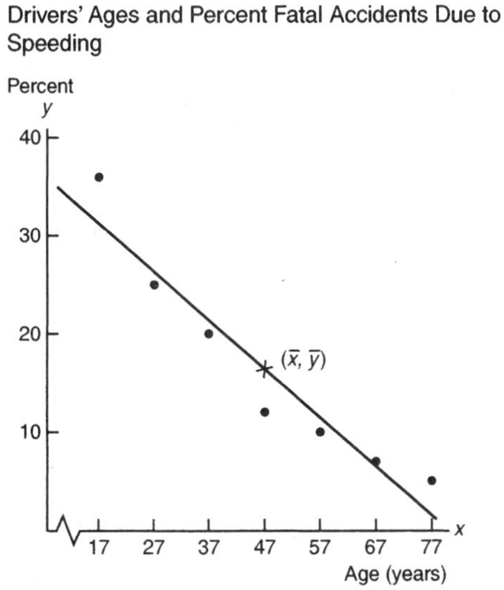

(b) Use a calculator to verify.

(c) $\bar{x} = \frac{\sum x}{n} = \frac{329}{7} = 47$ years

$\bar{y} = \frac{\sum y}{n} = \frac{115}{7} \approx 16.43\%$

$b = \frac{n\sum xy - (\sum x)(\sum y)}{n\sum x^2 - (\sum x)^2} = \frac{7(4015) - (329)(115)}{7(18,263) - (329)^2} \approx -0.496$

$a = \bar{y} - b\bar{x} \approx 16.43 - (-0.496)(47) \approx 39.761$

$\hat{y} = a + bx$ or $\hat{y} = 39.761 - 0.496x$

(d) See figure of part (a).

(e) $r^2 = (-0.959)^2 \approx 0.920$

This means that 92% of the variation in y = percentage of all fatal accidents due to speeding can be explained by the corresponding variation in x = age in years of a licensed automobile driver using the least squares line. 100% – 92% = 8% of the variation is unexplained.

(f) Use $x = 25$.

$\hat{y} = 39.761 - 0.496(25) \approx 27.36\%$

7. (a)

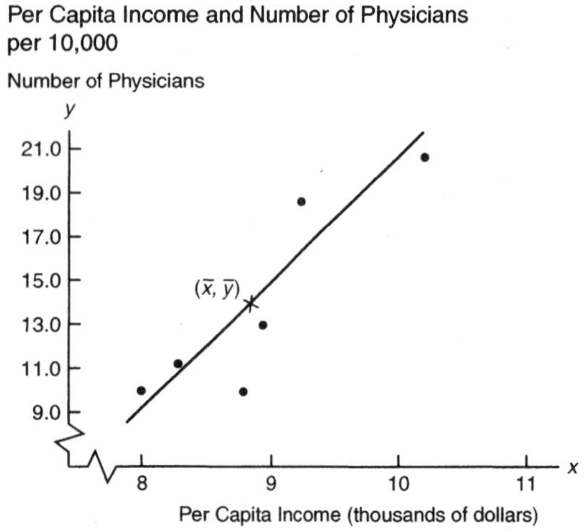

(b) Use a calculator to verify.

(c) $\bar{x} = \frac{\sum x}{n} = \frac{53}{6} \approx \$8.83$ thousand

$\bar{y} = \frac{\sum y}{n} = \frac{83.7}{6} = 13.95$ M.D.'s per 10,000

$b = \frac{n\sum xy - (\sum x)(\sum y)}{n\sum x^2 - (\sum x)^2} = \frac{6(755.89) - (53)(83.7)}{6(471.04) - (53)^2} \approx 5.756$

$a = \bar{y} - b\bar{x} \approx 13.95 - 5.756(8.83) \approx -36.898$

$\hat{y} = a + bx$ or $\hat{y} \approx -36.898 + 5.756x$

(d) See figure of part (a).

(e) $r^2 = (0.934)^2 \approx 0.872$

This means that 87.2% of the variation in y = number of medical doctors per 10,000 residents can be explained by the corresponding variation in x = per capita income in thousands of dollars using the least squares line. 100% – 87.2% = 12.8% of the variation is unexplained.

(f) Use $x = 10$.
$\hat{y} = -36.898 + 5.756(10) \approx 20.7$ M.D.'s per 10,000 residents

9. (a) Percentage of 16 to 19-Year-Olds Not in School and Number of Violent Crimes per 1000

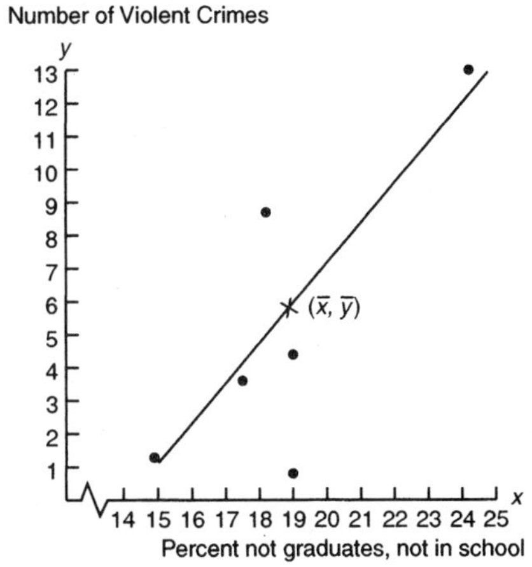

(b) Use a calculator to verify.

(c) $\bar{x} = \frac{\sum x}{n} = \frac{112.8}{6} = 18.8\%$

$\bar{y} = \frac{\sum y}{n} = \frac{32.4}{6} = 5.4$ crimes per 1000

$b = \frac{n\sum xy - (\sum x)(\sum y)}{n\sum x^2 - (\sum x)^2} = \frac{6(665.03) - (112.8)(32.4)}{6(2167.14) - (112.8)^2} \approx 1.202$

$a = \bar{y} - b\bar{x} \approx 5.4 - 1.202(18.8) \approx -17.204$

$\hat{y} = a + bx$ or $\hat{y} = -17.204 + 1.202x$

(d) See figure of part (a).

(e) $r^2 = (0.764)^2 \approx 0.584$

This means that 58.4% of the variation in y = reported violent crimes per 1000 residents can be explained by the corresponding variation in x = percentage of 16-to-19-year-olds not in school and not high school graduates using the least squares line. 100% – 58.4% = 41.6% of the variation is unexplained.

(f) Use $x = 24$.
$\hat{y} = -17.204 + 1.202(24) \approx 11.6$ crimes per 1000

11. (a) Cultural Affiliation and Elevation of Archaeological Sites

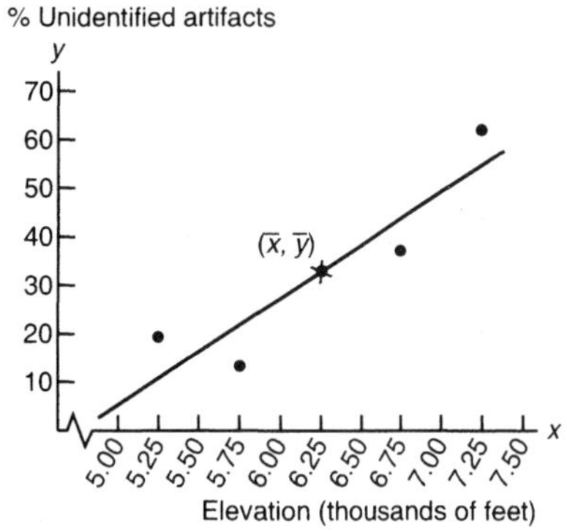

(b) Use a calculator to verify.

(c) $\bar{x} = \frac{\sum x}{n} = \frac{31.25}{5} = 6.25$

$\bar{y} = \frac{\sum y}{n} = \frac{164}{5} = 32.8$

$b = \frac{n\sum xy - (\sum x)(\sum y)}{n\sum x^2 - (\sum x)^2} = \frac{5(1080) - (31.25)(164)}{5(197.813) - (31.25)^2} = 22.0$

$a = \bar{y} - b\bar{x} = 32.8 - 22.0(6.25) = -104.5$

$\hat{y} = a + bx$ or $\hat{y} = -104.7 + 22.0x$

(d) See figure of part (a).

(e) $r^2 = (0.913)^2 \approx 0.833$

This means that 83.3% of the variation in y = % unidentified artifacts can be explained by the corresponding variation in x = elevation. 100% – 83.3% = 16.7% of the variation is unexplained.

(f) Use $x = 6.5$.

$\hat{y} = -104.7 + 22.0(6.5) = 38.3$ percent

13. (a) Elevation and the Number of Frost-Free Days

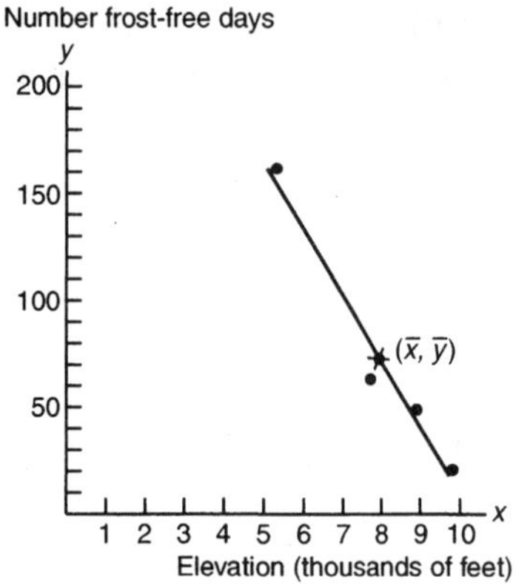

(b) Use a calculator to verify.

(c) $\overline{x} = \frac{\sum x}{n} = \frac{39.6}{5} = 7.92$

$\overline{y} = \frac{\sum y}{n} = \frac{368}{5} = 73.6$

$b = \frac{n\sum xy - (\sum x)(\sum y)}{n\sum x^2 - (\sum x)^2} = \frac{5(2562.3) - (39.6)(368)}{5(325.04) - (39.6)^2} \approx -30.88$

$a = \overline{y} - b\overline{x} = 73.6 - (-30.88)(7.92) \approx 318.16$

$\hat{y} = a + bx$ or $\hat{y} = 318.16 - 30.88x$

(d) See figure of part (a).

(e) $r^2 = (-0.981)^2 \approx 0.962$

This means that 96.2% of the variation in y = number of days can be explained by the corresponding variation in x = elevation using the least squares line. 100% – 92.6% = 3.8% of the variation is unexplained.

(f) Use $x = 6$.

$\hat{y} = 318.16 - 30.878(6) = 132.89$ days

15. (a)

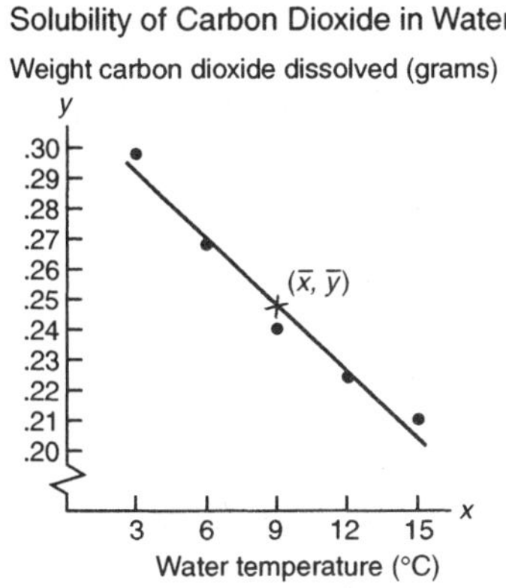

(b) Use a calculator to verify.

(c) $\overline{x} = \frac{\sum x}{n} = \frac{45}{5} = 9.0$

$\overline{y} = \frac{\sum y}{n} = \frac{1.24}{5} = 0.248$

$b = \frac{n\sum xy - (\sum x)(\sum y)}{n\sum x^2 - (\sum x)^2} = \frac{5(10.5) - (45)(1.24)}{5(495) - (45)^2} \approx -0.00733$

$a = \overline{y} - b\overline{x} = 0.248 - (-0.00733)(9.0) = 0.314$

$\hat{y} = a + bx$ or $\hat{y} = 0.314 - 0.00733x$

(d) See figure of part (a).

(e) $r^2 = (-0.985)^2 \approx 0.970$

This means that 97.0% of the variation in y = weight can be explained by the corresponding variation in x = temperature using the least squares line. 100% – 97.0% = 3% of the variation is unexplained.

(f) Use $x = 10$.

$\hat{y} = 0.314 - 0.00733(10) = 0.241$ gram

17. (a) Yes. The pattern of residuals appears randomly scattered around the horizontal line at 0.

(b) No. There do not appear to be any outliers.

19. (a) Result checks.

(b) Result checks.

(c) Yes.

(d) $y = 0.143 + 1.071x$

$$y - 0.143 = 1.071x$$

$$\frac{y - 0.143}{1.071} = x$$

$$\frac{1}{1.071}y - \frac{0.143}{1.071} = x$$

or

$x = 0.9337y - 0.1335$

The equation $x = 0.9337y - 0.1335$ does not match part (b), with the symbols x and y exchanged.

(e) In general, switching x and y values produces a different least-squares equation. It is important that when you perform a linear regression, you know which variable is the explanatory variable and which is the response variable.

Section 10.3

1. (a) Use a calculator to verify.

(b) $\alpha = 0.05$

$H_0: \rho = 0$

$H_1: \rho > 0$

$$t = \frac{r\sqrt{n-2}}{\sqrt{1-r^2}} = \frac{0.784\sqrt{6-2}}{\sqrt{1-0.784^2}} \approx 2.526$$

$d.f. = n - 2 = 6 - 2 = 4$

From Table 6, 2.526 falls between entries 2.132 and 2.776. Use the one-tailed areas to find that $0.025 < P\text{-value} < 0.050$.

Since the P-value interval ≤ 0.05 for α, we reject H_0.

At the 5% level of significance, there seems to be a positive correlation between x and y.

(c) Use a calculator to verify.

(d) $\hat{y} = a + bx$ or $\hat{y} = 16.542 + 0.4117x$

Use $x = 70$.

$\hat{y} = 16.542 + 0.4117(70) \approx 45.36\%$

(e) $d.f. = 4$, $t_c = 2.132$

$$E = t_c S_e \sqrt{1 + \frac{1}{n} + \frac{n(x-\bar{x})^2}{n\sum x^2 - (\sum x)^2}} = 2.132(2.6964)\sqrt{1 + \frac{1}{6} + \frac{6(70-73.167)^2}{6(32.393) - (439)^2}} \approx 6.31$$

The 90% confidence interval is

$$\hat{y} - E \le y \le \hat{y} + E$$

$$45.36 - 6.31 \le y \le 4536 + 6.31$$

$$39.05 \le y \le 51.67$$

(f) $\alpha = 0.05$

$H_0: \beta = 0$

$H_1: \beta > 0$

$$t = \frac{b}{S_e}\sqrt{\sum x^2 - \frac{1}{n}\left(\sum x\right)^2} = \frac{0.4117}{2.6964}\sqrt{32{,}393 - \frac{1}{6}(439)^2} \approx 2.522$$

$d.f. = n - 2 = 6 - 2 = 4$

From Table 6, 2.522 falls between entries 2.132 and 2.776. Use the one-tailed areas to find that $0.025 < P\text{-value} < 0.050$.

Since the P-value interval < 0.05 for α, we reject H_0.

At the 5% level of significance, there seems to be a positive slope between x and y.

(g) $d.f. = 4,\ t_c = 2.132,\ b \approx 0.4117$

$$E = \frac{t_c S_e}{\sqrt{\sum x^2 - \frac{1}{n}\left(\sum x\right)^2}} \approx \frac{2.132(2.6964)}{\sqrt{32{,}393 - \frac{1}{6}(439)^2}} \approx 0.3480$$

The 90% confidence interval is

$$b - E < \beta < b + E$$
$$0.4117 - 0.3480 < \beta < 0.4117 + 0.3480$$
$$0.064 < \beta < 0.760$$

For every percentage increase in successful free throws, the percentage of successful field goals increases by an amount between 0.06 to 0.76.

3. (a) Use a calculator to verify.

(b) $\alpha = 0.01$

$H_0: \rho = 0$

$H_1: \rho < 0$

$$t = \frac{r\sqrt{n-2}}{\sqrt{1-r^2}} = \frac{-0.976\sqrt{7-2}}{\sqrt{1-(-0.976)^2}} = -10.02$$

$d.f. = n - 2 = 7 - 2 = 5$

From Table 6, 10.02 falls to the right of entry 6.869. Use the one-tailed areas to find $P\text{-value} < 0.0005$.

Since the P-value interval ≤ 0.01 for α, we reject H_0.

At the 1% level of significance, the evidence supports a negative correlation between x and y.

(c) Use a calculator to verify.

(d) $\hat{y} = a + bx$ or $\hat{y} = 3.366 - 0.0544x$

Use $x = 18$.

$\hat{y} = 3.366 - 0.0544(18) \approx 2.39$ hours

(e) $d.f. = 5,\ t_c = 1.476,\ \bar{x} = \frac{\sum x}{n} = \frac{201.4}{7} \approx 28.77$

$$E = t_c S_e\sqrt{1 + \frac{1}{n} + \frac{n(x-\bar{x})^2}{n\sum x^2 - \left(\sum x\right)^2}} = 1.476(0.1660)\sqrt{1 + \frac{1}{7} + \frac{7(18-28.77)^2}{7(6735.46) - (201.4)^2}} \approx 0.276$$

The 80% confidence interval is

$$\hat{y} - E \le y \le \hat{y} + E$$
$$2.39 - 0.276 \le y \le 2.39 + 0.276$$
$$2.11 \le y \le 2.67$$

(f) $\alpha = 0.01$

$H_0: \beta = 0$

$H_1: \beta < 0$

$$t = \frac{b}{S_e}\sqrt{\sum x^2 - \frac{1}{n}\left(\sum x\right)^2} = \frac{-0.0544}{0.1660}\sqrt{6735.46 - \frac{1}{7}(201.4)^2} \approx -10.02$$

From Table 6, 10.02 falls to the right of entry 6.869. Use the one-tailed areas to find *P*-value < 0.0005.

Since the *P*-value interval ≤ 0.01 for α, we reject H_0.

At the 1% level of significance, the sample evidence supports a negative slope.

(g) $d.f. = 5,\ t_c = 2.015,\ b \approx -0.0544$

$$E = \frac{t_c S_e}{\sqrt{\sum x^2 - \frac{1}{n}\left(\sum x\right)^2}} = \frac{2.015(0.1660)}{\sqrt{6735.46 - \frac{1}{7}(201.4)^2}} \approx 0.011$$

The 90% confidence interval is

$$b - E < \beta < b + E$$
$$-0.054 - 0.011 < \beta < -0.054 + 0.011$$
$$-0.065 < \beta < -0.043$$

For every meter more of depth, the optimal time decreases from about 0.04 to 0.07 hour.

5. (a) Use a calculator to verify.

(b) $\alpha = 0.01$

$H_0: \rho = 0$

$H_1: \rho > 0$

$$t = \frac{r\sqrt{n-2}}{\sqrt{1-r^2}} = \frac{0.956\sqrt{6-2}}{\sqrt{1-(0.956)^2}} = 6.517$$

$d.f. = n - 2 = 6 - 2 = 4$

From Table 6, the sample test statistic falls between entries 4.604 and 8.610. Use one-tailed areas to find that 0.0005 < *P*-value < 0.005.

Since the *P*-value interval ≤ 0.01 for α, reject H_0.

At the 1% level of significance, the sample evidence supports a positive correlation.

(c) Use a calculator to verify.

(d) $\hat{y} = a + bx$ or $\hat{y} = 1.965 + 0.7577x$

Use $x = 14$.

$\hat{y} = 1.965 + 0.7577(14) \approx \$12.57$ thousand.

(e) $d.f. = 4,\ t_c = 1.778,\ \bar{x} = \frac{\sum x}{n} = \frac{79.6}{6} \approx 13.267$

$$E = t_c S_e \sqrt{1 + \frac{1}{n} + \frac{n(x-\bar{x})^2}{n\sum x^2 - \left(\sum x\right)^2}} = 1.778(0.1527)\sqrt{1 + \frac{1}{6} + \frac{6(14-13.267)^2}{6(1057.8) - (79.6)^2}} \approx 0.329$$

The 85% confidence interval is

$$\hat{y} - E \le y \le \hat{y} + E$$
$$12.57 - 0.329 \le y \le 12.57 + 0.329$$
$$12.24 \le y \le 12.90$$

(f) $\alpha = 0.01$

$H_0: \beta = 0$

$H_1: \beta > 0$

$$t = \frac{b}{S_e}\sqrt{\sum x^2 - \frac{1}{n}\left(\sum x\right)^2} = \frac{0.7577}{0.1527}\sqrt{(1057.8) - \frac{1}{6}(79.6)^2} \approx 6.608$$

$d.f. = n - 2 = 6 - 2 = 4$

From Table 6, the sample test statistic falls between entries 4.604 and 8.610. Use one-tailed areas to find that $0.0005 < P\text{-value} < 0.005$.

Since the P-value interval ≤ 0.01 for α, reject H_0.

At the 1% level of significance, the sample evidence supports a positive slope.

(g) $d.f. = 4$, $t_c = 2.776$, $b \approx 0.7577$

$$E = \frac{t_c S_e}{\sqrt{\sum x^2 - \frac{1}{n}\left(\sum x\right)^2}} = \frac{2.776(0.1527)}{\sqrt{(1057.8) - \frac{(79.6)^2}{6}}} \approx 0.318$$

The 95% confidence interval is

$$b - E < \beta < b + E$$
$$0.758 - 0.318 < \beta < 0.758 + 0.318$$
$$0.44 < \beta < 1.08$$

For every \$1,000 increase in list price, the dealer price increases between \$440 and \$1080.

7. (a) $\alpha = 0.01$

$H_0: \rho = 0$

$H_1: \rho \neq 0$

$$t = \frac{r\sqrt{n-2}}{\sqrt{1-r^2}} = \frac{0.90\sqrt{6-2}}{\sqrt{1-(0.90)^2}} = 4.129$$

$d.f. = n - 2 = 6 - 2 = 4$

From Table 6, 4.129 falls between entries 3.747 and 4.604. Use two-tailed areas to find $0.010 < P\text{-value} < 0.020$. Since the P-value interval > 0.01 for α, we do not reject H_0. The correlation coefficient ρ is not significantly different from zero at the 1% level of significance.

(b) $\alpha = 0.01$

$H_0: \rho = 0$

$H_1: \rho \neq 0$

$$t = \frac{r\sqrt{n-2}}{\sqrt{1-r^2}} = \frac{0.90\sqrt{10-2}}{\sqrt{1-(0.90)^2}} = 5.840$$

$d.f. = n - 2 = 10 - 2 = 8$

From Table 6, 5.840 falls to the right of entry 5.041. Use two-tailed areas to find $P\text{-value} < 0.001$. Since the P-value ≤ 0.01 for α, we reject H_0. The correlation coefficient ρ is significantly different from zero at the 1% level of significance.

(c) From part (a) to part (b), n increased from 6 to 10, the test statistic t increased from 4.129 to 5.840. For the same $r = 0.90$ and the same level of significance $\alpha = 0.01$, we rejected H_0 for the larger n but not for the smaller n.

In general, as n increases, the degrees of freedom $(n - 2)$ increase and the test statistic $\left(t = \frac{r\sqrt{n-2}}{\sqrt{1-r^2}}\right)$ increases. This produces a smaller P-value.

Section 10.4

1. $x_1 = 1.6 + 3.5x_2 - 7.9x_3 + 2.0x_4$

(a) The response variable is x_1.

(b) The constant term is 1.6.
The coefficient 3.5 goes with corresponding explanatory variable x_2.
The coefficient –7.9 goes with corresponding explanatory variable x_3.
The coefficient 2.0 goes with corresponding explanatory variable x_4.

(c) $x_2 = 2, x_3 = 1, x_4 = 5$
$x_1 = 1.6 + 3.5(2) - 7.9(1) + 2.0(5) = 10.7$
The predicted value is 10.7.

(d) In multiple regression, the coefficients of the explanatory variables can be thought of as "slopes" (the change in the response variables per unit change in the explanatory variable) if we look at one explanatory variable's coefficient at a time, while holding the other explanatory variables as arbitrary and fixed constants.
x_3 and x_4 held constant, x_2 increased by one unit:
The change in x_1 would be an increase of 3.5 units.
x_3 and x_4 held constant, x_2 increased by two units:
The change in x_1 would be an increase of 2(3.5) = 7 units.
x_3 and x_4 held constant, x_2 decreased by four units:
The change in x_1 would be an decrease of 4(3.5) = 14 units.

(e) $d.f. = n - k - 1 = 12 - 3 - 1 = 8$
A 90% confidence interval for the coefficient of x_2 is $b_2 - tS_2 < \beta_2 < b_2 + tS_2$
$3.5 - 1.86(0.419) < \beta_2 < 3.5 + 1.86(0.419)$
$2.72 < \beta_2 < 4.28$

(f) $\alpha = 0.05$
$H_0\colon \beta_2 = 0$
$H_1\colon \beta_2 \neq 0$

$$t = \frac{b_2 - \beta_2}{S_2} = \frac{3.5 - 0}{0.419} = 8.35$$

$d.f. = 8$
From Table 6, 8.35 falls to the right of entry 5.041. Use two-tailed areas to find that P-value < 0.001.
Since $0.001 \leq 0.05$ for α, reject H_0.
We conclude that $\beta_2 \neq 0$ and x_2 should be included as an explanatory variable in the least-squares equation.

3. (a)

	$\overline{x}$	s	$CV = \frac{s}{\overline{x}} \cdot 100$
x_1	150.09	13.63	9.08%
x_2	62.45	9.11	14.59%
x_3	195.0	17.31	8.88%

Relative to its mean, x_2 has the greatest spread of data values and x_3 has the smallest spread of data values.

(b) $r^2_{x_1x_2} \approx (0.979)^2 \approx 0.958$
$r^2_{x_1x_3} \approx (0.971)^2 \approx 0.943$
$r^2_{x_2x_3} \approx (0.946)^2 \approx 0.895$

The variable x_2 has the greatest influence on x_1 $(0.958 > 0.943)$.

Yes. Both variables x_2 and x_3 show a strong influence on x_1 because 0.958 and 0.943 are close to 1.

95.8% of the variation of x_1 can be explained by the corresponding variation in x_2.

94.3% of the variation of x_1 can be explained by the corresponding variation in x_3.

(c) $R^2 = 0.977$

97.7% of the variation in x_1 can be explained by the corresponding variation in x_2 and x_3 taken together.

(d) $x_1 = 30.99 + 0.861x_2 + 0.355x_3$

In multiple regression, the coefficients of the explanatory variables can be thought of as "slopes" (the change in the response variables per unit change in the explanatory variable) if we look at one explanatory variable's coefficient at a time, while holding the other explanatory variables as arbitrary and fixed constants.

If age (x_2) were held fixed and x_3 increased by 10 pounds, the systolic blood pressure is expected to increase by 0.335(10) = 3.35.

If weight (x_3) were held fixed and x_2 increased by 10 years, the systolic blood pressure is expected to increase by 0.861(10) = 8.61.

(e) $\alpha = 0.05$
H_0: $\beta_i = 0$
H_1: $\beta_i \neq 0$
$d.f = n - k - 1 = 11 - 2 - 1 = 8$

For β_2, the sample test statistic is $t = 3.47$ with P-value = 0.008.

For β_3, the sample test statistic is $t = 2.56$ with P-value = 0.034.

Since $0.008 \leq 0.05$ and $0.034 \leq 0.05$, reject H_0 for each coefficient and conclude that the coefficients of x_2 and x_3 are not zero. Explanatory variables x_i whose coefficients (β_i) are nonzero contribute information in the least squares equation, i.e., without these x_i, the resulting least squares regression equation is not as good a fit to the data as is the regression equation which includes these x_i.

(f) $d.f. = 8, t = 1.86$

A 90% confidence interval for β_i is

$$b_i - tS_i < \beta_i < b_i + tS_i$$
$$0.861 - 1.86(0.2482) < \beta_2 < 0.861 + 1.86(0.2482)$$
$$0.40 < \beta_2 < 1.32$$
$$0.335 - 1.86(0.1307) < \beta_3 < 0.335 + 1.86(0.1307)$$
$$0.09 < \beta_3 < 0.58$$

(g) $x_1 = 30.99 + 0.861(68) + 0.335(192) \approx 153.9$

Michael's predicted systolic blood pressure is 153.9 and a 90% confidence interval for this new observation's value, given these x_i, i.e., the prediction interval, is 148.3 to 159.4.

5. (a)

	$\bar{x}$	s	$CV = \frac{s}{\bar{x}} \cdot 100$
x_1	85.24	33.79	39.64%
x_2	8.74	3.89	44.51%
x_3	4.90	2.48	50.61%
x_4	9.92	5.17	52.12%

Relative to its mean, x_4 has the largest spread of data values. The larger the CV, the more we expect the variable to change relative to its average value, because a variable with a large CV has a large standard deviation, s, relative to $\bar{x}$, and s measures "spread," or variability, in the data. x_1 has a small CV because we divide by a large mean.

(b) $r^2_{x_1x_2} \approx (0.917)^2 \approx 0.841$

$r^2_{x_1x_3} \approx (0.930)^2 \approx 0.865$

$r^2_{x_1x_4} \approx (0.475)^2 \approx 0.226$

$r^2_{x_2x_3} \approx (0.790)^2 \approx 0.624$

$r^2_{x_2x_4} \approx (0.429)^2 \approx 0.184$

$r^2_{x_3x_4} \approx (0.299)^2 \approx 0.089$

The variable x_4 has the least influence on box office receipts x_1 $(0.226 < 0.841 < 0.865)$.

x_2 = production costs, $r^2_{x_1x_2} \approx 0.841$.

84.1% of the variation of box office receipts can be attributed to the corresponding variation in production costs.

(c) $R^2 = 0.967$

96.7% of the variation in x_1 can be explained by the corresponding variation in x_2, x_3, and x_4 taken together.

(d) $x_1 = 7.68 + 3.66x_2 + 7.62x_3 + 0.83x_4$

In multiple regression, the coefficients of the explanatory variables can be thought of as "slopes" (the change in the response variables per unit change in the explanatory variable) if we look at one explanatory variable's coefficient at a time, while holding the other explanatory variables as arbitrary and fixed constants.

If x_2 and x_4 were held fixed and x_3 were increased by 1 ($1 million), the corresponding change in x_1 (box office receipts) would be an increase of 7.62 or 7.62 million dollars.

(e) $\alpha = 0.05$
$H_0: \beta_i = 0$
$H_1: \beta_i \neq 0$
$d.f = n - k - 1 = 10 - 3 - 1 = 6$
For β_2, the sample test statistic is $t = 3.28$ with P-value = 0.017.
For β_3, the sample test statistic is $t = 4.60$ with P-value = 0.004.
For β_4, the sample test statistic is $t = 1.54$ with P-value = 0.175.
Since $0.017 \leq 0.05$ and $0.004 \leq 0.05$, reject H_0 for coefficients β_2 and β_3 and conclude that the coefficients of x_2 and x_3 are not zero.
Since $0.175 > 0.05$, do not reject H_0 for the coefficient β_4 and conclude that the coefficient of x_4 could be zero. If $\beta_4 = 0$, then x_4 contributes nothing to the (population) regression line. We can eliminate the variable x_4 and fit the (estimated) regression line without it, and probably see little, if any, difference between the predicted values of x_1 based on x_2 and x_3 only and the predicted values of x_1 based on x_2, x_3, and x_4.

(f) $d.f. = 6$, $t = 1.943$
A 90% confidence interval for β_i is

$$b_i - tS_i < \beta_i < b_i + tS_i$$

$$3.662 - 1.943(1.118) < \beta_2 < 3.662 + 1.943(1.118)$$
$$1.49 < \beta_2 < 5.83$$

$$7.621 - 1.943(1.657) < \beta_3 < 7.621 + 1.943(1.657)$$
$$4.40 < \beta_3 < 10.84$$

$$0.8285 - 1.943(0.5394) < \beta_4 < 0.8285 + 1.943(0.5394)$$
$$-0.22 < \beta_4 < 1.88$$

(g) $x_1 = 7.68 + 3.66(11.4) + 7.62(4.7) + 0.83(8.1) = 91.94$
The prediction is 91.94 million dollars and an 85% confidence interval for this new observation's value is \$77.6 million to \$106.3 million, given these x_i.

(h) $x_3 = -0.650 + 0.102x_1 - 0.260x_2 - 0.0899x_4$
$x_3 = -0.650 + 0.102(100) - 0.260(12) - 0.0899(9.2)$
$x_3 = 5.63$
The prediction is 5.63 million dollars and a 80% confidence interval for this new observation's value is \$4.21 million to \$7.04 million, given these x_i.

Chapter 10 Review

1. (a)

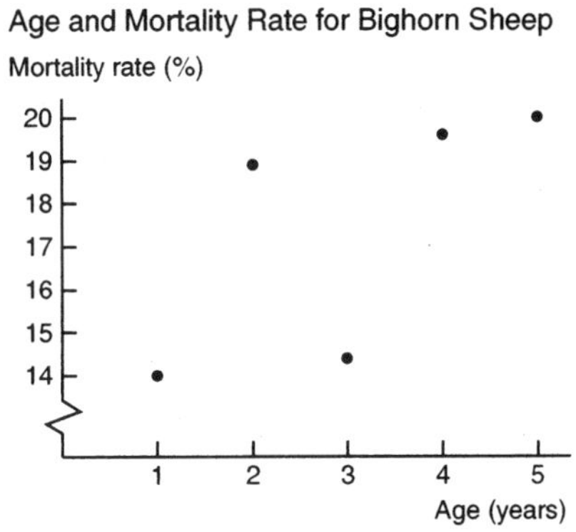

(b) $\overline{x} = \frac{\sum x}{n} = \frac{15}{5} = 3$

$\overline{y} = \frac{\sum y}{n} = \frac{86.9}{5} = 17.38$

$b = \frac{n\sum xy - (\sum x)(\sum y)}{n\sum x^2 - (\sum x)^2} = \frac{5(273.4) - (15)(86.9)}{5(55) - (15)^2} = 1.27$

$a = \overline{y} - b\overline{x} = 17.38 - 1.27(3) = 13.57$

$y = a + bx$ or $y = 13.57 + 1.27x$

(c) $r = \frac{n\sum xy - (\sum x)(\sum y)}{\sqrt{n\sum x^2 - (\sum x)^2}\sqrt{n\sum y^2 - (\sum y)^2}} = \frac{5(273.4) - (15)(86.9)}{\sqrt{5(55) - (15)^2}\sqrt{5(1544.73) - (86.9)^2}} \approx 0.685$

$r^2 = (0.685)^2 \approx 0.469$

The correlation coefficient r measures the strength of the linear relationship between a bighorn sheep's age and the mortality rate. The coefficient of determination r^2 means that 46.9% of the variation in y = mortality rate in this age groups can be explained by the corresponding variation in x = age of a bighorn sheep using the least squares line.

(d) $\alpha = 0.01$

$H_0\!: \rho = 0$

$H_1\!: \rho > 0$

$t = \frac{r\sqrt{n-2}}{\sqrt{1-r^2}} = \frac{0.685\sqrt{5-2}}{\sqrt{1-(0.685)^2}} = 1.629$

$d.f. = n - 2 = 5 - 2 = 3$

From Table 6, 1.629 falls between entries 1.423 and 1.638. Use one-tailed areas to find that $0.100 < P\text{-value} < 0.125$.

Since the P-value interval is > 0.01 for α, we do not reject H_0.

There does not seem to be a positive correlation between age and mortality rate of bighorn sheep.

(e) No, based on this limited data, predictions from the least-squares line model might be misleading. There appear to be other lurking variables that affect the mortality rate of sheep in different age groups.

3. (a)

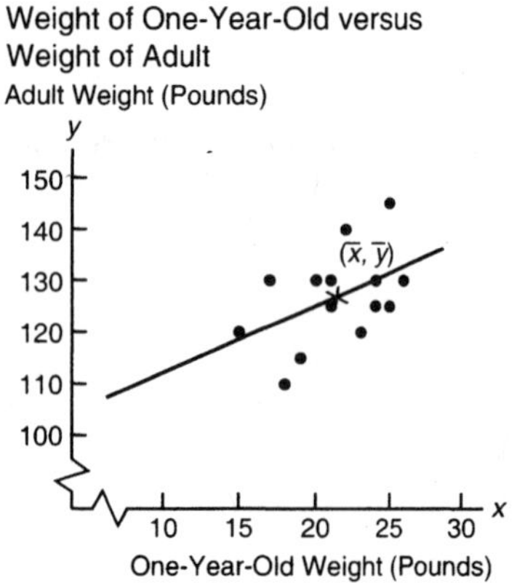

(b) $\overline{x} = \frac{\sum x}{n} = \frac{300}{14} = 21.43$

$\overline{y} = \frac{\sum y}{n} = \frac{1775}{14} = 126.79$

$b = \frac{n\sum xy - (\sum x)(\sum y)}{n\sum x^2 - (\sum x)^2} = \frac{14(38,220) - (300)(1775)}{14(6572) - (300)^2} \approx 1.285$

$a = \overline{y} - b\overline{x} = 126.79 - (1.285)(21.43) = 99.25$

$\hat{y} = a + bx$ or $\hat{y} = 99.25 + 1.285x$

See the figure in part (a).

(c) $r = \frac{n\sum xy - (\sum x)(\sum y)}{\sqrt{n\sum x^2 - (\sum x)^2}\sqrt{n\sum y^2 - (\sum y)^2}} = \frac{14(38,220) - (300)(1775)}{\sqrt{14(6572) - (300)^2}\sqrt{14(226,125) - (1775)^2}} \approx 0.468$

$r^2 = (0.468)^2 \approx 0.219$

The coefficient of determination r^2 means that 21.9% of the variation in y = weight of a mature adult (30 years old) can be explained by the corresponding variation in x = weight of a 1-year-old baby using the least squares line.

(d) $\alpha = 0.01$

$H_0: \rho = 0$

$H_1: \rho > 0$

$t = \frac{r\sqrt{n-2}}{\sqrt{1-r^2}} = \frac{0.468\sqrt{14-2}}{\sqrt{1-(0.468)^2}} = 1.834$

$d.f. = n - 2 = 14 - 2 = 12$

From Table 6, the sample test statistic falls between entries 1.782 and 2.179. Use one-tailed areas to find that $0.025 < P\text{-value} < 0.050$. Since the P-value interval > 0.01 for α, do not reject H_0. At the 1% level of significance, there does not seem to be a positive correlation between weight of baby and weight of adult.

(e) Let $x = 20$.

$\hat{y} = 99.25 + 1.285(20) = 124.95$

The predicted weight is 124.95 pounds.

(f) $S_e = \sqrt{\frac{\sum y^2 - a\sum y - b\sum xy}{n-2}} = \sqrt{\frac{226,125 - 99.25(1775) - 1.285(38,220)}{14-2}} \approx 8.38$

(g) $E = t_c S_e \sqrt{1 + \frac{1}{n} + \frac{n(x-\overline{x})^2}{n\sum x^2 - (\sum x)^2}}$

$= 2.179(8.38)\sqrt{1 + \frac{1}{14} + \frac{14(20-21.43)^2}{14(6572) - (300)^2}}$

≈ 19.03

A 95% confidence interval for y is

$$\hat{y} - E \le y \le \hat{y} + E$$
$$124.95 - 19.03 \le y \le 124.95 + 19.03$$
$$105.92 \le y \le 143.98$$

(h) $\alpha = 0.01$

$H_0\colon \beta = 0$

$H_1\colon \beta > 0$

$$t = \frac{b}{S_e}\sqrt{\sum x^2 - \frac{1}{n}\left(\sum x\right)^2} = \frac{1.285}{8.38}\sqrt{6572 - \frac{1}{14}(300)^2} \approx 1.84$$

$d.f. = n - 2 = 14 - 2 = 12$

From Table 6, the sample test statistic falls between entries 1.782 and 2.179. Use one-tailed areas to find that $0.025 < P\text{-value} < 0.050$. Since the P-value interval > 0.01 for α, do not reject H_0. At the 1% level of significance, there does not seem to be a positive slope between weight of baby and weight of adult.

(i) $d.f. = 12,\ t_c = 1.356,\ b = 1.285$

$$E = \frac{t_c S_e}{\sqrt{\sum x^2 - \frac{\left(\sum x\right)^2}{n}}} = \frac{1.356(8.38)}{\sqrt{6572 - \frac{(300)^2}{14}}} \approx 0.949$$

An 80% confidence interval is

$$b - E < \beta < b + E$$

$$1.285 - 0.949 < \beta < 1.285 + 0.949$$

$$0.34 < \beta < 2.22$$

At the 80% confidence level, we can say that for each additional pound a female infant weighs at one year, the adult weight changes by 0.34 to 2.22 lb.

5. (a) Pounds of Mail versus Number of Employees

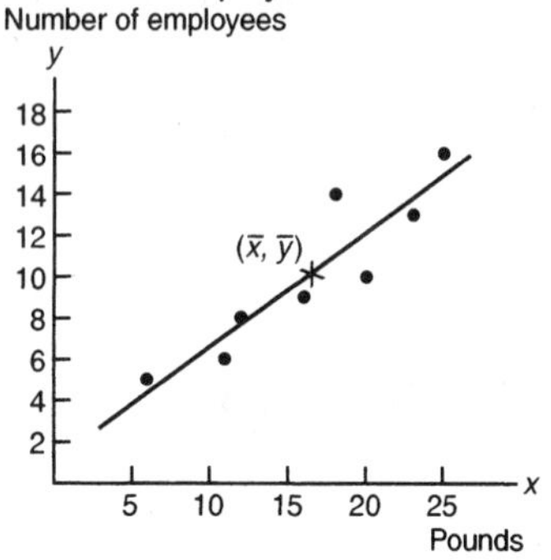

(b) $\overline{x} = \frac{\sum x}{n} = \frac{131}{8} = 16.375 \approx 16.38$

$\overline{y} = \frac{\sum y}{n} = \frac{81}{8} = 10.125 \approx 10.13$

$$b = \frac{n\sum xy - \left(\sum x\right)\left(\sum y\right)}{n\sum x^2 - \left(\sum x\right)^2} = \frac{8(1487) - (131)(81)}{8(2435) - (131)^2} \approx 0.554118 \approx 0.554$$

$a = \overline{y} - b\overline{x} = 10.125 - 0.554118(16.375) = 1.051$

$\hat{y} = a + bx$ or $\hat{y} = 1.051 + 0.554x$

See the figure in part (a).

(c) $r = \dfrac{n\sum xy - (\sum x)(\sum y)}{\sqrt{n\sum x^2 - (\sum x)^2}\sqrt{n\sum y^2 - (\sum y)^2}} = \dfrac{8(1487) - (131)(81)}{\sqrt{8(2435) - (131)^2}\sqrt{8(927) - (81)^2}} \approx 0.913$

$r^2 = (0.913)^2 \approx 0.833$

The coefficient of determination r^2 means that 83.3% of the variation in y = number of employees can be explained by the corresponding variation in x = weight of incoming mail using the least squares line.

(d) $\alpha = 0.01$

$H_0: \rho = 0$

$H_1: \rho > 0$

$t = \dfrac{r\sqrt{n-2}}{\sqrt{1-r^2}} = \dfrac{0.913\sqrt{8-2}}{\sqrt{1-(0.913)^2}} \approx 5.48$

$d.f. = n - 2 = 8 - 2 = 6$

From Table 6, the sample test statistic falls between entries 3.707 and 5.959. Use one-tailed areas to find that $0.0005 < P\text{-value} < 0.005$. Since the P-value interval ≤ 0.01 for α, reject H_0. At the 1% level of significance, there is sufficient evidence to show a positive correlation between pounds of mail and number of employees required to process the mail.

(e) Use $x = 15$.

$\hat{y} = 1.051 + 0.554(15) = 9.36$

About 9 or 10 employees should be assigned mail duty.

(f) $S_e = \sqrt{\dfrac{\sum y^2 - a\sum y - b\sum xy}{n-2}} = \sqrt{\dfrac{927 - 1.051(81) - 0.554118(1487)}{8-2}} \approx 1.73$

(g) $E = t_c S_e \sqrt{1 + \dfrac{1}{n} + \dfrac{n(x-\bar{x})^2}{n\sum x^2 - (\sum x)^2}}$

$= 2.447(1.73)\sqrt{1 + \dfrac{1}{8} + \dfrac{8(15 - 16.375)^2}{8(2435) - (131)^2}}$

≈ 4.5

A 95% confidence interval for y is

$$\hat{y} - E \le y \le \hat{y} + E$$
$$9.36 - 4.5 \le y \le 9.36 + 4.5$$
$$4.86 \le y \le 13.86$$

(h) $\alpha = 0.01$

$H_0: \beta = 0$

$H_1: \beta > 0$

$t = \dfrac{b}{S_e}\sqrt{\sum x^2 - \dfrac{1}{n}(\sum x)^2} = \dfrac{0.554118}{1.73}\sqrt{2435 - \dfrac{1}{8}(131)^2} \approx 5.45$

$d.f. = n - 2 = 8 - 2 = 6$

From Table 6, the sample test statistic falls between entries 3.707 and 5.959. Use one-tailed areas to find that $0.0005 < P\text{-value} < 0.005$. Since the P-value interval ≤ 0.01 for α, reject H_0. At the 1% level of significance, there is sufficient evidence to show a positive correlation between pounds of mail and number of employees required to process the mail.

(i) $d.f. = 6, t_c = 1.440, b \approx 0.554$

$$E = \frac{t_c S_e}{\sqrt{\sum x^2 - \frac{(\sum x)^2}{n}}} = \frac{1.440(1.73)}{\sqrt{2435 - \frac{(131)^2}{8}}} \approx 0.146$$

An 80% confidence interval is

$$b - E < \beta < b + E$$
$$0.554 - 0.146 < \beta < 0.554 + 0.146$$
$$0.41 < \beta < 0.70$$

At the 80% confidence level, we can say that for each additional pound of mail, between 0.4 and 0.7 additional employees are needed.

Chapter 11 Chi-Square and F Distributions

Section 11.1

1. (a) $\alpha = 0.05$
H_0: Myers-Briggs preference and profession are independent.
H_1: Myers-Briggs preference and profession are not independent.

(b) $\chi^2 = \sum \frac{(O-E)^2}{E}$

$$= \frac{(62-49.02)^2}{49.02} + \frac{(45-57.98)^2}{57.98} + \frac{(68-74.22)^2}{74.22} + \frac{(94-87.78)^2}{87.78} + \frac{(56-62.76)^2}{62.76} + \frac{(81-74.24)^2}{74.24}$$

$= 8.649$

All expected frequencies are greater than 5. Use the chi-square distribution.
Since there are 3 rows and 2 columns, $d.f. = (3-1)(2-1) = 2$.

(c) In Table 7, with $d.f. = 2$, $\chi^2 = 8.649$ falls between entries 7.38 and 9.21. Therefore, $0.010 < P\text{-value} < 0.025$.

(d) Since the P-value is less than the level of significance $\alpha = 0.05$, we reject the null hypothesis.

(e) At the 5% level of significance, there is sufficient evidence to conclude that Myers-Briggs preference and the profession are not independent.

3. (a) $\alpha = 0.01$
H_0: Site type and pottery type are independent.
H_1: Site type and pottery type are not independent.

(b) $\chi^2 = \sum \frac{(O-E)^2}{E}$

$$= \frac{(75-74.64)^2}{74.64} + \frac{(61-59.89)^2}{59.89} + \frac{(53-54.47)^2}{54.47} + \frac{(81-84.11)^2}{84.11} + \frac{(70-67.5)^2}{67.5}$$
$$+ \frac{(62-61.39)^2}{61.39} + \frac{(92-89.25)^2}{89.25} + \frac{(68-71.61)^2}{71.61} + \frac{(66-65.14)^2}{65.14}$$

$= 0.5552$

All expected frequencies are greater than 5. Use the chi-square distribution.
Since there are 3 rows and 3 columns, $d.f. = (3-1)(3-1) = 4$.

(c) In Table 7, with $d.f. = 4$, $\chi^2 = 0.5552$ falls between entries 0.484 and 0.711.
Therefore, $0.950 < P\text{-value} < 0.975$.

(d) Since the P-value is greater than the level of significance $\alpha = 0.01$, we do not reject the null hypothesis.

(e) At the 1% level of significance, there is insufficient evidence to conclude that site and pottery type are not independent.

5. (a) $\alpha = 0.05$

H_0: Age distribution and location are independent.

H_1: Age distribution and location are not independent.

(b) $\chi^2 = \sum \frac{(O-E)^2}{E}$

$$= \frac{(13-14.08)^2}{14.08} + \frac{(13-12.84)^2}{12.84} + \frac{(15-14.08)^2}{14.08} + \frac{(10-11.33)^2}{11.33} + \frac{(11-10.34)^2}{10.34} + \frac{(12-11.33)^2}{11.33} + \frac{(34-31.59)^2}{31.59} + \frac{(28-28.82)^2}{28.82} + \frac{(30-31.59)^2}{31.59}$$

$$= 0.67$$

All expected frequencies are greater than 5. Use the chi-square distribution.
Since there are 3 rows and 3 columns, $d.f. = (3-1)(3-1) = 4$.

(c) In Table 7, with $d.f. = 4$, $\chi^2 = 0.67$ falls between entries 0.484 and 0.711.
Therefore, $0.950 < P\text{-value} < 0.975$.

(d) Since the P-value is greater than the level of significance $\alpha = 0.05$, we do not reject the null hypothesis.

(e) At the 5% level of significance, there is insufficient evidence to conclude that age distribution and location are not independent.

7. (a) $\alpha = 0.05$

H_0: Ages of young adult and movie preference are independent.

H_1: Ages of young adult and movie preference are not independent.

(b) $\chi^2 = \sum \frac{(O-E)^2}{E}$

$$= \frac{(8-10.60)^2}{10.60} + \frac{(15-12.06)^2}{12.06} + \frac{(11-11.33)^2}{11.33} + \frac{(12-9.35)^2}{9.35} + \frac{(10-10.65)^2}{10.65} + \frac{(8-10.00)^2}{10.00} + \frac{(9-9.04)^2}{9.04} + \frac{(8-10.29)^2}{10.29} + \frac{(12-9.67)^2}{9.67}$$

$$= 3.623$$

All expected frequencies are greater than 5. Use the chi-square distribution.
Since there are 3 rows and 3 columns, $d.f. = (3-1)(3-1) = 4$.

(c) In Table 7, with $d.f. = 4$, $\chi^2 = 3.623$ falls between entries 1.064 and 7.78.
Therefore, $0.100 < P\text{-value} < 0.900$.

(d) Since the P-value is greater than the level of significance $\alpha = 0.05$, we do not reject the null hypothesis.

(e) At the 5% level of significance, there is insufficient evidence to conclude that age of young adult and movie preference are not independent.

9. (a) $\alpha = 0.05$

H_0: Ticket sales and type of billing are independent.

H_1: Ticket sales and type of billing are not independent.

(b) $\chi^2 = \sum \frac{(O-E)^2}{E}$

$$= \frac{(10-7.52)^2}{7.52} + \frac{(12-13.16)^2}{13.16} + \frac{(18-18.80)^2}{18.80} + \frac{(7-7.52)^2}{7.52}$$
$$+ \frac{(6-8.48)^2}{8.48} + \frac{(16-14.84)^2}{14.84} + \frac{(22-21.20)^2}{21.20} + \frac{(9-8.48)^2}{8.48}$$
$$= 1.87$$

All expected frequencies are greater than 5. Use the chi-square distribution.
Since there are 2 rows and 4 columns, $d.f. = (2-1)(4-1) = 3$.

(c) In Table 7, with $d.f. = 3$, $\chi^2 = 1.87$ falls between entries 0.584 and 6.25.
Therefore, $0.100 < P\text{-value} < 0.900$.

(d) Since the P-value is greater than the level of significance $\alpha = 0.05$, we do not reject the null hypothesis.

(e) At the 5% level of significance, there is insufficient evidence to conclude that ticket sales and type of billing are not independent.

11. (a) $\alpha = 0.05$

H_0: Stone tool construction material and site are independent.

H_1: Stone tool construction material and site are not independent.

(b) $\chi^2 = \sum \frac{(O-E)^2}{E}$

$$= \frac{(731-689.36)^2}{689.36} + \frac{(584-625.64)^2}{625.64} + \frac{(102-102.22)^2}{102.22} + \frac{(93-92.78)^2}{92.78}$$
$$+ \frac{(510-542.58)^2}{542.58} + \frac{(525-492.42)^2}{492.42} + \frac{(85-93.84)^2}{93.84} + \frac{(94-85.16)^2}{85.16}$$
$$= 11.15$$

All expected frequencies are greater than 5. Use the chi-square distribution.
Since there are 4 rows and 2 columns, $d.f. = (4-1)(2-1) = 3$.

(c) In Table 7, with $d.f. = 3$, $\chi^2 = 11.15$ falls between entries 9.35 and 11.34.
Therefore, $0.010 < P\text{-value} < 0.025$.

(d) Since the P-value is less than the level of significance $\alpha = 0.05$, we reject the null hypothesis.

(e) At the 5% level of significance, there is sufficient evidence to conclude that stone tool construction material and site are not independent.

Section 11.2

1. (a) $\alpha = 0.05$

H_0: The distributions are the same.

H_1: The distributions are different.

(b) $\chi^2 = \sum \frac{(O-E)^2}{E}$

$$= \frac{(47-32.76)^2}{32.76} + \frac{(75-61.88)^2}{61.88} + \frac{(288-305.31)^2}{305.31} + \frac{(45-55.06)^2}{55.06}$$

$= 11.79$

All expected frequencies are greater than 5. Use the chi-square distribution.
$d.f. = k - 1 = 4 - 1 = 3$.

(c) In Table 7, with $d.f. = 3$, $\chi^2 = 11.79$ falls between entries 11.34 and 12.84.
Therefore, $0.005 < P\text{-value} < 0.010$.

(d) Since the P-value is less than the level of significance $\alpha = 0.05$, we reject the null hypothesis.

(e) At the 5% level of significance, the evidence is sufficient to conclude that the Red Lake Village population does not fit the general Canadian population.

3. (a) $\alpha = 0.01$

H_0: The distributions are the same.

H_1: The distributions are different.

(b) $\chi^2 = \sum \frac{(O-E)^2}{E}$

$$= \frac{(906-910.92)^2}{910.92} + \frac{(162-157.52)^2}{157.52} + \frac{(168-169.40)^2}{169.40} + \frac{(197-194.67)^2}{194.67} + \frac{(53-53.50)^2}{53.50}$$

$= 0.1984$

All expected frequencies are greater than 5. Use the chi-square distribution.
$d.f. = k - 1 = 5 - 1 = 4$.

(c) In Table 7, with $d.f. = 4$, the entry 0.207 has 0.995 area in the right tail. Note that as χ^2 decreases, the area in the right tail increases so a $\chi^2 < 0.207$ means the corresponding $P\text{-value} > 0.995$.

(d) Since the P-value is greater than the level of significance $\alpha = 0.01$, we do not reject the null hypothesis.

(e) At the 1% level of significance, the evidence is insufficient to conclude that the regional distribution of raw materials does not fit the distribution at the current excavation site.

5. (i) Essay.

(ii) (a) H_0: The distributions are the same.
H_1: The distributions are different.

(b) $\alpha = 0.01$

$$\chi^2 = \sum \frac{(O-E)^2}{E}$$
$$= \frac{(16-14.57)^2}{14.57} + \frac{(78-83.70)^2}{83.70} + \frac{(212-210.80)^2}{210.80} + \frac{(221-210.80)^2}{210.80} + \frac{(81-83.70)^2}{83.70} + \frac{(12-14.57)^2}{14.57}$$
$$= 1.5693$$

All expected frequencies are greater than 5. Use the chi-square distribution.
$d.f. = k - 1 = 6 - 1 = 5$.

(c) In Table 7, with $d.f. = 5$, $\chi^2 = 1.5693$ falls between entries 1.145 and 1.61. Therefore, $0.900 < P\text{-value} < 0.950$.

(d) Since the P-value is greater than the level of significance $\alpha = 0.01$, we do not reject the null hypothesis.

(e) At the 1% level of significance, the evidence is insufficient to conclude that the average daily July temperature does not follow a normal distribution.

7. (a) $\alpha = 0.05$
H_0: The distributions are the same.
H_1: The distributions are different.

(b) $\alpha = 0.05$

$$\chi^2 = \sum \frac{(O-E)^2}{E}$$
$$= \frac{(120-150)^2}{150} + \frac{(85-75)^2}{75} + \frac{(220-200)^2}{200} + \frac{(75-75)^2}{75}$$
$$= 9.333$$

All expected frequencies are greater than 5. Use the chi-square distribution.
$d.f. = k - 1 = 4 - 1 = 3$.

(c) In Table 7, with $d.f. = 3$, $\chi^2 = 9.333$ falls between entries 7.81 and 9.35. Therefore, $0.025 < P\text{-value} < 0.050$.

(d) Since the P-value is less than the level of significance $\alpha = 0.05$, we reject the null hypothesis.

(e) At the 5% level of significance, the evidence is sufficient to conclude that current fish distribution is different from that of five years ago.

9. (a) $\alpha = 0.01$
H_0: The distributions are the same.
H_1: The distributions are different.

(b) $\chi^2 = \sum \frac{(O-E)^2}{E}$

$$= \frac{(127-121.50)^2}{121.50} + \frac{(40-36.45)^2}{36.45} + \frac{(480-461.70)^2}{461.70} + \frac{(502-498.15)^2}{498.15} + \frac{(56-72.90)^2}{72.90} + \frac{(10-24.30)^2}{24.30}$$

$= 13.7$

All expected frequencies are greater than 5. Use the chi-square distribution.
$d.f. = k - 1 = 6 - 1 = 5$.

(c) In Table 7, with $d.f. = 5$, $\chi^2 = 13.7$ falls between entries 12.83 and 15.09. Therefore, $0.010 < P\text{-value} < 0.025$.

(d) Since the P-value is greater than the level of significance $\alpha = 0.01$, we do not reject the null hypothesis.

(e) At the 1% level of significance, the evidence is insufficient to conclude that census distribution and the ethnic origin distribution of city residents are different.

11. (a) $\alpha = 0.01$
H_0: The distributions are the same.
H_1: The distributions are different.

(b) $\chi^2 = \sum \frac{(O-E)^2}{E}$

$$= \frac{(83-82.775)^2}{82.775} + \frac{(49-48.4)^2}{48.4} + \frac{(32-34.375)^2}{34.375} + \frac{(22-26.675)^2}{26.675} + \frac{(25-21.725)^2}{21.725} + \frac{(18-18.425)^2}{18.425} + \frac{(13-15.95)^2}{15.95} + \frac{(17-14.025)^2}{14.025} + \frac{(16-12.65)^2}{12.65}$$

$= 3.559$

All expected frequencies are greater than 5. Use the chi-square distribution.
$d.f. = k - 1 = 9 - 1 = 8$.

(c) In Table 7, with $d.f. = 8$, $\chi^2 = 3.559$ falls between entries 3.49 and 13.36. Therefore, $0.100 < P\text{-value} < 0.900$.

(d) Since the P-value interval is greater than the level of significance $\alpha = 0.01$, we do not reject the null hypothesis.

(e) At the 1% level of significance, the evidence is insufficient to conclude that distribution of first-non-zero digits in the accounting file do not follow Benford's Law.

13. (a) $P(r) = \dfrac{e^{-\lambda}\lambda^r}{r!}$

$$P(0) = \frac{e^{-1.72}(1.72)^0}{0!} \approx 0.179$$

$$P(1) = \frac{e^{-1.72}(1.72)^1}{1!} \approx 0.308$$

$$P(2) = \frac{e^{-1.72}(1.72)^2}{2!} \approx 0.265$$

$$P(3) = \frac{e^{1.72}(1.72)^3}{3!} \approx 0.152$$

$$\begin{aligned} P(4 \text{ or more}) &= 1 - P(3 \text{ or less}) \\ &= 1 - (0.179 + 0.308 + 0.265 + 0.152) \\ &= 0.096 \end{aligned}$$

(b)

r	$E = 90P(r)$
0	90(0.179) = 16.11
1	90(0.308) = 27.72
2	90(0.265) = 23.85
3	90(0.152) = 13.68
4 or more	90(0.096) = 8.64

(c) $$\begin{aligned} \chi^2 &= \sum \frac{(O-E)^2}{E} \\ &= \frac{(22-16.11)^2}{16.11} + \frac{(21-27.72)^2}{27.72} + \frac{(15-23.85)^2}{23.85} \\ &\quad + \frac{(17-13.68)^2}{13.68} + \frac{(15-8.64)^2}{8.64} \\ &= 12.55 \end{aligned}$$

$d.f. = k - 1 = 5 - 1 = 4$

(d) $\alpha = 0.01$

H_0: The Poisson distribution fits.

H_1: The Poisson distribution does not fit.

In Table 7, with $d.f. = 4$, $\chi^2 = 12.55$ falls between entries 11.14 and 13.28. Therefore, $0.010 < P\text{-value} < 0.025$. Since the P-value is greater than the level of significance $\alpha = 0.01$, we do not reject the null hypothesis. At the 1% level of significance, we cannot say that the Poisson distribution does not fit.

Section 11.3

1. (a) $\alpha = 0.05$

H_0: $\sigma^2 = 42.3$

H_1: $\sigma^2 > 42.3$

(b) $\chi^2 = \frac{(n-1)s^2}{\sigma^2} = \frac{(23-1)46.1}{42.3} = 23.98$

$d.f. = 23 - 1 = 22$

Assume a normal population distribution.

(c) Since > is in H_1, a right-tailed test is used.

Using Table 7 and $d.f. = 22$, $\chi^2 = 23.98$ falls between entries 14.04 and 30.81. Therefore, $0.100 < P\text{-value} < 0.900$.

(d) Since the P-value is greater than the level of significance $\alpha = 0.05$, we do not reject the null hypothesis.

(e) At the 5% level of significance, there is insufficient evidence to conclude the variance is greater in the new section.

(f) For $d.f. = 22$ and $\alpha = \frac{1-0.95}{2} = 0.025$, $\chi^2_U = 36.78$.

For $d.f. = 22$ and $\alpha = \frac{1+0.95}{2} = 0.975$, $\chi^2_L = 10.98$.

The 95% confidence interval for σ^2 is

$$\frac{(n-1)s^2}{\chi^2_U} < \sigma^2 < \frac{(n-1)s^2}{\chi^2_L}$$

$$\frac{(23-1)46.1}{36.78} < \sigma^2 < \frac{(23-1)46.1}{10.98}$$

$$27.57 < \sigma^2 < 92.37$$

3. (a) $\alpha = 0.01$

H_0: $\sigma^2 = 136.2$

H_1: $\sigma^2 < 136.2$

(b) $\chi^2 = \frac{(n-1)s^2}{\sigma^2} = \frac{(8-1)115.1}{136.2} = 5.92$

$d.f. = 8 - 1 = 7$

Assume a normal population distribution.

(c) Since < is in H_1, a left-tailed test is used.

Using Table 7 and $d.f. = 7$, $\chi^2 = 5.92$ falls between entries 2.83 and 12.02. Therefore,

$0.900 < \text{Right-tail area} < 0.100$

$1 - 0.900 < P\text{-value} < 1 - 0.100$

$0.100 < P\text{-value} < 0.900$

(d) Since the P-value is greater than the level of significance, we do not reject the null hypothesis.

(e) At the 1% level of significance, there is insufficient evidence to conclude the variance for number of mountain-climber deaths is less than 136.2

(f) For $d.f. = 7$ and $\alpha = \frac{1-0.90}{2} = 0.05$, $\chi_U^2 = 14.07$.

For $d.f. = 7$ and $\alpha = \frac{1+0.90}{2} = 0.95$, $\chi_L^2 = 2.17$.

The 90% confidence interval for σ^2 is

$$\frac{(n-1)s^2}{\chi_U^2} < \sigma^2 < \frac{(n-1)s^2}{\chi_L^2}$$

$$\frac{(8-1)115.1}{14.07} < \sigma^2 < \frac{(8-1)115.1}{2.17}$$

$$57.26 < \sigma^2 < 371.29$$

5. (a) $\alpha = 0.05$
H_0: $\sigma^2 = 9$
H_1: $\sigma^2 < 9$

(b) $\chi^2 = \frac{(n-1)s^2}{\sigma^2} = \frac{(23-1)(1.9)^2}{3^2} = 8.82$
$d.f. = 23 - 1 = 22$
Assume a normal population distribution.

(c) Since < is in H_1, a left-tailed test is used.
Using Table 7 and $d.f. = 22$, $\chi^2 = 8.82$ falls between entries 8.64 and 9.54. Therefore,
$0.995 < \text{Right-tail area} < 0.990$
$1 - 0.995 < P\text{-value} < 1 - 0.990$
$0.005 < P\text{-value} < 0.010$

(d) Since the *P*-value is less than the level of significance, we reject the null hypothesis.

(e) At the 5% level of significance, there is sufficient evidence to conclude the variance of protection times for the new typhoid shot is less than 9.

(f) For $d.f. = 22$ and $\alpha = \frac{1-0.90}{2} = 0.05$, $\chi_U^2 = 33.92$.

For $d.f. = 22$ and $\alpha = \frac{1+0.90}{2} = 0.95$, $\chi_L^2 = 12.34$.

The 90% confidence interval for σ is

$$\sqrt{\frac{(n-1)s^2}{\chi_U^2}} < \sigma < \sqrt{\frac{(n-1)s^2}{\chi_L^2}}$$

$$\sqrt{\frac{(23-1)(1.9)^2}{33.92}} < \sigma < \sqrt{\frac{(23-1)(1.9)^2}{12.34}}$$

$$1.53 < \sigma < 2.54$$

7. (a) $\alpha = 0.01$
H_0: $\sigma^2 = 0.18$
H_1: $\sigma^2 > 0.18$

(b) $\chi^2 = \dfrac{(n-1)s^2}{\sigma^2} = \dfrac{(61-1)0.27}{0.18} = 90$

$d.f. = 61 - 1 = 60$

Assume a normal population distribution.

(c) Since > is in H_1, a right-tailed test is used.

Using Table 7 and $d.f. = 60$, $\chi^2 = 90$ falls between entries 88.38 and 91.95. Therefore, $0.005 < P\text{-value} < 0.010$.

(d) Since the *P*-value is less than the level of significance, we reject the null hypothesis.

(e) At the 1% level of significance, there is sufficient evidence to conclude the variance of measurements on the fan blades is higher than the specified amount. The inspector is justified in claiming the blades must be replaced.

(f) For $d.f. = 60$ and $\alpha = \dfrac{1-0.90}{2} = 0.05$, $\chi_U^2 = 79.08$.

For $d.f. = 60$ and $\alpha = \dfrac{1+0.90}{2} = 0.95$, $\chi_L^2 = 43.19$.

The 90% confidence interval for σ is

$$\sqrt{\frac{(n-1)s^2}{\chi_U^2}} < \sigma < \sqrt{\frac{(n-1)s^2}{\chi_L^2}}$$

$$\sqrt{\frac{(61-1)0.27}{79.08}} < \sigma < \sqrt{\frac{(61-1)0.27}{43.19}}$$

$$0.45 \text{ mm} < \sigma < 0.61 \text{ mm}$$

9. (i) **(a)** $\alpha = 0.05$

H_0: $\sigma^2 = 23$

H_1: $\sigma^2 \neq 23$

(b) $\chi^2 = \dfrac{(n-1)s^2}{\sigma^2} = \dfrac{(22-1)(14.3)}{23} = 13.06$

$d.f. = 22 - 1 = 21$

Assume a normal population distribution.

(c) The area to the left of $\chi^2 = 13.8$ is less than 50%, so we double the left-tail area to find the *P*-value for the two-tailed test. Right-tail area is between 0.950 and 0.900. Subtracting each value from 1, we find the left-tail area is between 0.050 and 0.100. Doubling the left-tail area for a two-tailed test gives $0.100 < P\text{-value} < 0.200$.

(d) Since the *P*-value is greater than the level of significance $\alpha = 0.05$, we do not reject the null hypothesis.

(e) At the 5% level of significance, there is insufficient evidence to conclude the variance of battery life is different from 19.

(ii) For $d.f. = 21$ and $\alpha = \frac{1-0.90}{2} = 0.05$, $\chi_U^2 = 32.67$.

For $d.f. = 21$ and $\alpha = \frac{1+0.90}{2} = 0.95$, $\chi_L^2 = 11.59$.

The 90% confidence interval for σ^2 is

$$\frac{(n-1)s^2}{\chi_U^2} < \sigma^2 < \frac{(n-1)s^2}{\chi_L^2}$$

$$\frac{(22-1)(14.3)}{32.67} < \sigma^2 < \frac{(22-1)(14.3)}{11.59}$$

$$9.19 < \sigma^2 < 25.91$$

(iii) The 90% confidence interval for σ is

$$\sqrt{\frac{(n-1)s^2}{\chi_U^2}} < \sigma < \sqrt{\frac{(n-1)s^2}{\chi_L^2}}$$

$$\sqrt{9.19} < \sigma < \sqrt{25.91}$$

$$3.03 < \sigma < 5.09$$

Section 11.4

1. (a) $\alpha = 0.01$; Population I is annual production from the first plot.

H_0: $\sigma_1^2 = \sigma_2^2$

H_1: $\sigma_1^2 > \sigma_2^2$

(b) Since $s^2 \approx 0.332$ is larger than $s^2 \approx 0.089$, we designate Population I as the first plot.

$$F = \frac{s_1^2}{s_2^2} = \frac{0.332}{0.089} = 3.73$$

$d.f._N = n_1 - 1 = 16 - 1 = 15$, and $d.f._D = n_2 - 1 = 16 - 1 = 15$

The populations follow independent normal distributions. The samples are random samples from each population.

(c) Since > is in H_1, a right-tailed test is used.

From Table 8, $F = 3.73$ falls between entries 3.52 and 5.54. Therefore, $0.001 < P\text{-value} < 0.010$.

(d) Since the P-value is less than the level of significance $\alpha = 0.01$, we reject the null hypothesis.

(e) At the 1% level of significance, there is sufficient evidence to show that the variance in annual wheat production from the first plot is greater than that of the second plot.

3. (a) $\alpha = 0.05$; Population I has data from France.

H_0: $\sigma_1^2 = \sigma_2^2$

H_1: $\sigma_1^2 \neq \sigma_2^2$

(b) Since $s^2 \approx 2.044$ is larger than $s^2 \approx 1.038$, we designate Population I as France.

$$F = \frac{s_1^2}{s_2^2} = \frac{2.044}{1.038} = 1.97$$

$d.f._N = n_1 - 1 = 21 - 1 = 20$, and $d.f._D = n_2 - 1 = 18 - 1 = 17$
The populations follow independent normal distributions. The samples are random samples from each population.

(c) Since $\neq$ is in H_1, a two-tailed test is used.
From Table 8, $F = 1.97$ falls between entries 1.86 and 2.23. Therefore,
$0.050 <$ Right-tail area < 0.100
$0.100 < P$-value < 0.200

(d) Since the P-value is greater than the level of significance $\alpha = 0.05$, we do not reject the null hypothesis.

(e) At the 5% level of significance, there is insufficient evidence to show that the variance in corporate productivity of large companies in France and that of those in Germany differ. Volatility of corporate productivity does not appear to differ.

5. (a) $\alpha = 0.05$; Population I has data from aggressive growth companies.
H_0: $\sigma_1^2 = \sigma_2^2$
H_1: $\sigma_1^2 > \sigma_2^2$

(b) Since $s^2 \approx 348.43$ is larger than $s^2 \approx 137.31$, we designate Population I as aggressive growth.

$$F = \frac{s_1^2}{s_2^2} = \frac{348.43}{137.31} = 2.54$$

$d.f._N = n_1 - 1 = 21 - 1 = 20$, and $d.f._D = n_2 - 1 = 21 - 1 = 20$
The populations follow independent normal distributions. The samples are random samples from each population.

(c) Since $>$ is in H_1, a right-tailed test is used.
From Table 8, $F = 2.54$ falls between entries 2.46 and 2.94. Therefore,
$0.010 < P$-value < 0.025

(d) Since the P-value is less than the level of significance $\alpha = 0.05$, we reject the null hypothesis.

(e) At the 5% level of significance, there is sufficient evidence to show that the variance in percentage annual returns for funds holding aggressive-growth small stocks is larger than that for funds holding value stocks.

7. (a) $\alpha = 0.05$; Population I has data from the new system.
H_0: $\sigma_1^2 = \sigma_2^2$
H_1: $\sigma_1^2 \neq \sigma_2^2$

(b) Since $s^2 \approx 58.4$ is larger than $s^2 \approx 31.6$, we designate Population I as the new system.

$$F = \frac{s_1^2}{s_2^2} = \frac{58.4}{31.6} = 1.85$$

$d.f._N = n_1 - 1 = 31 - 1 = 30$, and $d.f._D = n_2 - 1 = 25 - 1 = 24$

The populations follow independent normal distributions. The samples are random samples from each population.

(c) Since $\neq$ is in H_1, a two-tailed test is used.

From Table 8, $F = 1.85$ falls between entries 1.67 and 1.94. Therefore,

$0.050 < \text{Right-tail area} < 0.100$

$0.100 < P\text{-value} < 0.200$

(d) Since the P-value is greater than the level of significance $\alpha = 0.05$, we do not reject the null hypothesis.

(e) At the 5% level of significance, there is insufficient evidence to show that the variance in gasoline consumption for the two injection systems is different.

Section 11.5

1. (a) $\alpha = 0.01$

H_0: $\mu_1 = \mu_2 = \mu_3$

H_1: Not all the means are equal.

(b)

Site I	Site II	Site III
$n = 7$	$n = 4$	$n = 6$
$\sum x_1 = 286$	$\sum x_2 = 164$	$\sum x_3 = 176$
$\sum x_1^2 = 15{,}312$	$\sum x_2^2 = 8354$	$\sum x_3^2 = 7450$
$SS_1 = 3626.857$	$SS_2 = 1630$	$SS_3 = 2287.33\overline{3}$

$$\sum x_{TOT} = 286 + 164 + 176 = 626$$

$$\sum x_{TOT}^2 = 15{,}312 + 8354 + 7450 = 31{,}116$$

$$N = 7 + 4 + 6 = 17$$

$$k = 3$$

$$SS_{TOT} = \sum x_{TOT}^2 - \frac{(\sum x_{TOT})^2}{N} = 31{,}116 - \frac{(626)^2}{17} = 8064.470$$

$$SS_{BET} = \sum_{\text{all groups}} \left(\frac{(\sum x_i)^2}{n_i} \right) - \frac{(\sum x_{TOT})^2}{N}$$

$$= \frac{(286)^2}{7} + \frac{(164)^2}{4} + \frac{(176)^2}{6} - \frac{(626)^2}{17} = 520.280$$

$$SS_W = SS_1 + SS_2 + SS_3 = 3626.857 + 1630 + 2287.333 = 7544.190$$

Check that $SS_{TOT} = SS_{BET} + SS_W$: $8064.470 = 520.280 + 7544.190$

$$d.f._{BET} = k - 1 = 3 - 1 = 2$$
$$d.f._W = N - k = 17 - 3 = 14$$
$$d.f._{TOT} = N - 1 = 17 - 1 = 16$$
$$MS_{BET} = \frac{SS_{BET}}{d.f._{BET}} = \frac{520.280}{2} = 260.14$$
$$MS_W = \frac{SS_W}{d.f._W} = \frac{7544.190}{14} = 538.87$$
$$F = \frac{MS_{BET}}{MS_W} = \frac{260.14}{538.87} = 0.48$$

For $d.f._N = 2$ and $d.f._D = 14$.

(c) From Table 8, $F = 0.48$ falls below the entry 2.73. Therefore, P-value > 0.100.

(d) Since the P-value is greater than the level of significance $\alpha = 0.01$, do not reject H_0.

(e) At the 1% level of significance, there is insufficient evidence to conclude the means are not all equal.

(f) Summary of ANOVA results

Source of Variation	Sum of Squares	Degrees of Freedom	MS	F Ratio	P-value	Test Decision
Between groups	520.280	2	260.14	0.48	> 0.100	Do not reject H_0
Within groups	7544.190	14	538.87			
Total	8064.470	16				

3. (a) $\alpha = 0.05$

$H_0: \mu_1 = \mu_2 = \mu_3 = \mu_4$

H_1: Not all the means are equal.

(b) See Section 11.5 or solutions to problems 1-2 for examples of calculations from formulas.

$$F = \frac{MS_{BET}}{MS_W} = \frac{29.879}{35.325} \approx 0.846$$

$d.f._N = 3$ and $d.f._D = 18$

(c) From Table 8, $F = 0.846$ falls below entry 2.42. Therefore, P-value > 0.100.

(d) Since the P-value is greater than the level of significance $\alpha = 0.05$, we do not reject H_0.

(e) At the 5% level of significance, there is insufficient evidence to conclude the means are not all equal.

(f) Summary of ANOVA results

Source of Variation	Sum of Squares	Degrees of Freedom	MS	F Ratio	P-value	Test Decision
Between groups	89.637	3	29.879	0.846	> 0.100	Do not reject H_0
Within groups	635.827	18	35.324			
Total	725.464	21				

5. (a) $\alpha = 0.05$
$H_0: \mu_1 = \mu_2 = \mu_3$
H_1: Not all the means are equal.

(b) See Section 11.5 or solutions to problems 1-2 for examples of calculations from formulas.

$$F = \frac{MS_{BET}}{MS_W} = \frac{651.58}{130.19} \approx 5.005$$
$d.f._N = 2$ and $d.f._D = 9$

(c) From Table 8, $F = 5.005$ falls between entries 4.26 and 5.71. Therefore, $0.025 < P\text{-value} < 0.050$.

(d) Since the P-value is less than the level of significance $\alpha = 0.05$, we reject H_0.

(e) At the 5% level of significance, there is sufficient evidence to conclude the means are not all equal.

(f) Summary of ANOVA results

Source of Variation	Sum of Squares	Degrees of Freedom	MS	F Ratio	P-value	Test Decision
Between groups	1303.167	2	651.58	5.005	$0.025 < P\text{-value} < 0.050$	Reject H_0
Within groups	1171.750	9	130.19			
Total	2474.917	11				

7. (a) $\alpha = 0.01$
$H_0: \mu_1 = \mu_2 = \mu_3$
H_1: Not all the means are equal.

(b) See Section 11.5 or solutions to problems 1-2 for examples of calculations from formulas.

$$F = \frac{MS_{BET}}{MS_W} = \frac{1.021}{3.039} \approx 0.336$$
$d.f._N = 2$ and $d.f._D = 11$

(c) From Table 8, $F = 0.336$ falls below entry 2.86. Therefore, $P\text{-value} > 0.100$.

(d) Since the P-value is greater than the level of significance $\alpha = 0.01$, we do not reject H_0.

(e) At the 1% level of significance, there is insufficient evidence to conclude the means are not all equal.

(f) Summary of ANOVA results

Source of Variation	Sum of Squares	Degrees of Freedom	MS	F Ratio	P-value	Test Decision
Between groups	2.042	2	1.021	0.336	> 0.100	Do not reject H_0
Within groups	33.428	11	3.039			
Total	35.470	13				

9. (a) $\alpha = 0.05$

H_0: $\mu_1 = \mu_2 = \mu_3 = \mu_4$

H_1: Not all the means are equal.

(b) See Section 11.5 or solutions to problems 1-2 for examples of calculations from formulas.

$$F = \frac{MS_{BET}}{MS_W} = \frac{79.408}{17.223} \approx 4.611$$

$d.f._N = 3$ and $d.f._D = 15$

(c) From Table 8, $F = 4.611$ falls between entries 4.15 and 5.42. Therefore, $0.010 < P\text{-value} < 0.025$.

(d) Since the P-value is less than the level of significance $\alpha = 0.05$, we reject H_0.

(e) At the 5% level of significance, there is sufficient evidence to conclude the means are not all equal.

(f) Summary of ANOVA results

Source of Variation	Sum of Squares	Degrees of Freedom	MS	F Ratio	P-value	Test Decision
Between groups	238.225	3	79.408	4.611	$0.010 < P\text{-value} < 0.025$	Reject H_0
Within groups	258.340	15	17.223			
Total	496.565	18				

Section 11.6

1. There are two factors. One factor is *walking device* with 3 levels and the other factor is *task* with two levels. The data table has 6 cells.

3. Since the P value is less than 0.01, there is a significant difference in mean cadence according to the factor *walking device used.* The critical value is $F_{0.01} = 6.01$. Since the sample $F = 30.94$ is greater than $F_{0.01}$, F lies in the critical region and we reject H_0.

5. (a) There are two factors. One factor is *income level* with 4 levels and the other factor is *media type* with 5 levels.

(b) For income,

H_0: There is no difference in population mean index based on income level.
H_1: At least two income levels have different population mean indices.

$$F_{\text{income}} = \frac{MS_{\text{income}}}{MS_{\text{error}}} = \frac{359}{130} \approx 2.77$$

P-value ≈ 0.088
At the 5% level of significance, do not reject H_0.
The data do not indicate any differences in population mean index according to income level.

(c) For media,

H_0: No difference in population mean index by media type.
H_1: At least two media types have different population mean indices.

$$F_{\text{media}} = \frac{MS_{\text{media}}}{MS_{\text{error}}} = \frac{4}{130} \approx 0.03$$

P-value ≈ 0.998
At the 5% level of significance, do not reject H_0.
The data do not indicate any differences in population mean index according to media type.

7.

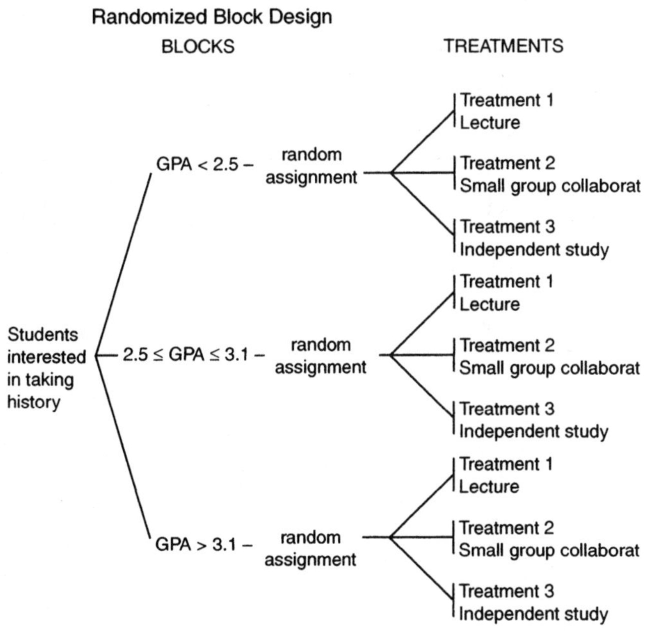

Yes, the design fits the model for randomized block design.

Chapter 11 Review

1. One-way ANOVA

(i) $\alpha = 0.01$
H_0: $\mu_1 = \mu_2 = \mu_3 = \mu_4$
H_1: Not all the means are equal.

(ii) See Section 11.5 or solutions to problems 1-2 for examples of calculations from formulas.

$$F = \frac{MS_{BET}}{MS_n} = \frac{2050}{778} \approx 2.63$$

$d.f._N = 3$ and $d.f._D = 16$

(iii) From Table 8, $F = 2.63$ falls between entries 2.46 and 3.24. Therefore, $0.050 < P\text{-value} < 0.100$.

(iv) Since the P-value is greater than the level of significance $\alpha = 0.01$, we do not reject H_0.

(v) At the 1% level of significance, there is insufficient evidence to conclude that not all the packaging mean sales are equal.

(vi) Summary of ANOVA results

Source of Variation	Sum of Squares	Degrees of Freedom	MS	F Ratio	P-value	Test Decision
Between groups	6150	3	2050	2.63	$0.050 < P\text{-value} < 0.100$	Do not reject H_0
Within groups	12,455	16	778			
Total	18,605	19				

3. (a) Chi-square for testing σ^2

(i) $\alpha = 0.01$
H_0: $\sigma^2 = 1{,}040{,}400$
H_1: $\sigma^2 > 1{,}040{,}400$

(ii) $$\chi^2 = \frac{(n-1)s^2}{\sigma^2} = \frac{(30-1)(1353)^2}{(1020)^2} = 51.03$$

$d.f. = 30 - 1 = 29$

(iii) Since > is in H_1, a right-tailed test is used.
In Table 7, with $d.f. = 29$, $\chi^2 = 51.03$ falls between entries 49.59 and 52.34. Therefore, $0.005 < P\text{-value} < 0.010$.

(iv) Since the P-value is less than the level of significance $\alpha = 0.01$, we reject the null hypothesis.

(v) At the 1% level of significance, there is sufficient evidence to conclude that the variance is greater than claimed.

(b) For $d.f. = 29$ and $\alpha = \dfrac{1-0.95}{2} = 0.025$, $\chi_U^2 = 45.72$.

For $d.f. = 29$ and $\alpha = \dfrac{1+0.95}{2} = 0.975$, $\chi_L^2 = 16.05$.

The 95% confidence interval for σ^2 is

$$\frac{(n-1)s^2}{\chi_U^2} < \sigma^2 < \frac{(n-1)s^2}{\chi_L^2}$$

$$\frac{(30-1)(1353)^2}{45.72} < \sigma^2 < \frac{(30-1)(1353)^2}{16.05}$$

$$1{,}161{,}147.4 < \sigma^2 < 3{,}307{,}642.4 \text{ square foot-pounds}$$

5. Chi-square test of independence

(i) $\alpha = 0.01$
H_0: Student grade and teacher rating are independent.
H_1: Student grade and teacher rating are not independent.

(ii) $\chi^2 = \sum \dfrac{(O-E)^2}{E}$

$$= \frac{(14-10.00)^2}{10.00} + \frac{(18-13.33)^2}{13.33} + \frac{(15-21.67)^2}{21.67} + \frac{(3-5.00)^2}{5.00} + \frac{(25-30.00)^2}{30.00}$$

$$+ \frac{(35-40.00)^2}{40.00} + \frac{(75-65.00)^2}{65.00} + \frac{(15-15.00)^2}{15.00} + \frac{(21-20.00)^2}{20.00} + \frac{(27-26.67)^2}{26.67}$$

$$+ \frac{(40-43.33)^2}{43.33} + \frac{(12-10.00)^2}{10.00}$$

$$= 9.8$$

Since there are 3 rows and 4 columns, $d.f. = (3-1)(4-1) = 6$.

(iii) In Table 7, with $d.f. = 6$, $\chi^2 = 9.8$ falls between entries 2.20 and 10.64. Therefore, $0.100 < P\text{-value} < 0.900$.

(iv) Since the P-value is greater than the level of significance $\alpha = 0.01$, we do not reject the null hypothesis.

(v) At the 1% level of significance, there is insufficient evidence to claim that student grade and teacher rating are not independent.

7. Chi-square test of goodness of fit

(i) $\alpha = 0.01$
H_0: The distributions are the same.
H_1: The distributions are different.

(ii) $\chi^2 = \sum \frac{(O-E)^2}{E}$

$$= \frac{(26-42.0)^2}{42.0} + \frac{(27-31.5)^2}{31.5} + \frac{(69-63.0)^2}{63.0} + \frac{(68-52.5)^2}{52.5} + \frac{(20-21.0)^2}{21.0}$$

$= 11.93$

$d.f. = k - 1 = 5 - 1 = 4$

(iii) In Table 7, with 4 $d.f.$, $\chi^2 = 11.93$ falls between entries 11.14 and 13.28. Therefore, $0.010 < P\text{-value} < 0.025$.

(iv) Since the P-value is greater than the level of significance $\alpha = 0.01$, we do not reject H_0.

(v) At the 1% level of significance, there is insufficient evidence to claim that the age distribution of the population of Blue Valley has changed.

9. F test for the equality of two variances

(i) $\alpha = 0.05$

H_0: $\sigma_1^2 = \sigma_2^2$

H_1: $\sigma_1^2 > \sigma_2^2$

(ii) Since $s^2 = 135.24$ is larger than $s^2 = 51.87$, we designate Population I as the new process.

$$F = \frac{s_1^2}{s_2^2} = \frac{135.24}{51.87} = 2.61$$

$d.f._N = n_1 - 1 = 16 - 1 = 15$, and $d.f._D = n_2 - 1 = 18 - 1 = 17$

(iii) Since > is in H_1, a right-tailed test is used.
From Table 8, $F = 2.61$ falls between entries 2.31 and 2.72. Therefore,
$0.025 < P\text{-value} < 0.050$.

(iv) Since the P-value is less than the level of significance $\alpha = 0.05$, we reject the null hypothesis.

(v) At the 5% level of significance, there is sufficient evidence to show that the variance for the lifetimes of bulbs manufactured using the new process is larger than that of bulbs made by the old process.

Chapter 12 Nonparametric Statistics

Section 12.1

1.

Region	Modern %	Historic %	Sign of Difference
1	4.0	3.3	+
2	2.3	1.9	+
3	7.8	7.0	+
4	2.8	5.5	−
5	0.7	3.3	−
6	5.1	6.0	−
7	2.9	3.2	−
8	4.2	8.2	−
9	4.9	6.4	−
10	5.8	7.2	−
11	6.8	6.1	+
12	3.6	1.5	+
13	3.2	1.0	+
14	0.8	2.1	−
15	7.3	5.1	+

(a) $\alpha = 0.05$

H_0: Distributions are the same.

H_1: Distributions are different.

(b) $x = \dfrac{\text{number of plus signs}}{\text{total number of signs}} = \dfrac{7}{15} \approx 0.4667$

$$z = \frac{x - 0.5}{\sqrt{\frac{0.25}{n}}} = \frac{0.4667 - 0.5}{\sqrt{\frac{0.25}{15}}} \approx -0.26$$

Use the standard normal distribution.

(c) By Table 5, $P(z < -0.26) = 0.3974$. For a two-tailed test P-value $= 2(0.3974) = 0.7948$.

(d) Since the P-value is greater than the level of significance $\alpha = 0.05$, do not reject H_0.

(e) At the 5% level of significance, the data are not significant. The evidence is insufficient to conclude that the economies are different.

3.

Student	After	Before	Sign of Difference
1	107	111	–
2	115	110	+
3	120	93	+
4	78	75	+
5	83	88	–
6	56	56	N.D.
7	71	75	–
8	89	73	+
9	77	83	–
10	44	40	+
11	119	115	+
12	130	101	+
13	91	110	–
14	99	90	+
15	96	98	–
16	83	76	+
17	100	100	N.D.
18	118	109	+

(a) $\alpha = 0.05$

H_0: Distributions are the same.

H_1: Distributions are different.

(b) $x = \frac{\text{number of plus signs}}{\text{total number of signs}} = \frac{10}{16} = 0.625$

$z = \frac{x - 0.5}{\sqrt{\frac{0.25}{n}}} = \frac{0.625 - 0.5}{\sqrt{\frac{0.25}{16}}} = 1.00$

Use the standard normal distribution.

(c) By Table 5, $P(z > 1.00) = 0.1587$. For a two-tailed test, P-value = 2(0.1587) = 0.3174.

(d) Since the P-value is greater than the level of significance $\alpha = 0.05$, do not reject H_0.

(e) At the 5% level of significance, the data are not significant. The evidence is insufficient to conclude that the lectures have any effect on student awareness of current events.

5.

Twin Pair	A	B	Sign of Difference
1	177	86	+
2	150	135	+
3	112	115	–
4	95	110	–
5	120	116	+
6	117	84	+
7	86	93	–
8	111	77	+
9	110	96	+
10	142	130	+
11	125	147	–
12	89	101	–

(a) $\alpha = 0.05$

H_0: Distributions are the same.

H_1: Distributions are different.

(b) $x = \frac{\text{number of plus signs}}{\text{total number of signs}} = \frac{7}{12} \approx 0.5833$

$$z = \frac{x - 0.5}{\sqrt{\frac{0.25}{n}}} = \frac{0.5833 - 0.5}{\sqrt{\frac{0.25}{12}}} \approx 0.58$$

Use the standard normal distribution.

(c) By Table 5, $P(z > 0.58) = 0.2810$. For a two-tailed test, P-value = 2(0.2810) = 0.5620.

(d) Since the P-value is greater than the level of significance $\alpha = 0.05$, do not reject H_0.

(e) At the 5% level of significance, the data are not significant. The evidence is insufficient to conclude that the schools area not equally effective.

7.

Subject	After	Before	Sign of Difference
1	28	28	N.D.
2	15	35	–
3	2	14	–
4	20	20	N.D.
5	31	25	+
6	19	40	–
7	6	18	–
8	17	15	+
9	1	21	–
10	5	19	–
11	12	32	–
12	20	42	–
13	30	26	+
14	19	37	–
15	0	19	–
16	16	38	–
17	4	23	–
18	19	24	–

(a) $\alpha = 0.01$

H_0: Distributions are the same.

H_1: Distribution after hypnosis is lower.

(b) $x = \frac{\text{number of plus signs}}{\text{total number of signs}} = \frac{3}{16} = 0.1875$

$$z = \frac{x - 0.5}{\sqrt{\frac{0.25}{n}}} = \frac{0.1875 - 0.5}{\sqrt{\frac{0.25}{16}}} = -2.5$$

Use the standard normal distribution.

(c) By Table 5, the area in the left tail is $P(z < -2.5) = 0.0062$; P-value = 0.0062.

(d) Since the P-value is less than the level of significance $\alpha = 0.01$, reject H_0.

(e) At the 1% level of significance, the data are significant. The evidence is sufficient to conclude that the number of cigarettes smoked per day was less after hypnosis.

9.

Region	Male	Female	Sign of Difference
1	7.3	7.5	–
2	7.5	6.4	+
3	7.7	6.0	+
4	21.8	20.0	+
5	4.2	2.6	+
6	12.2	5.2	+
7	3.5	3.1	+
8	4.2	4.9	–
9	8.0	12.1	–
10	9.7	10.8	–
11	14.1	15.6	–
12	3.6	6.3	–
13	3.6	4.0	–
14	4.0	3.9	+
15	5.2	9.8	–
16	6.9	9.8	–
17	15.6	12.0	+
18	6.3	3.3	+
19	8.0	7.1	+
20	6.5	8.2	–

(a) $\alpha = 0.01$

H_0: Distributions are the same.

H_1: Distributions are different.

(b) $x = \frac{\text{number of plus signs}}{\text{total number of signs}} = \frac{10}{20} = 0.500$

$$z = \frac{x - 0.5}{\sqrt{\frac{0.25}{n}}} = \frac{0.500 - 0.5}{\sqrt{\frac{0.25}{20}}} = 0$$

Use the standard normal distribution.

(c) By Table 5, $P(z > 0) = 0.5000$. For a two-tailed test P-value $= 2(0.5000) = 1.000$.

(d) Since the P-value is greater than the level of significance $\alpha = 0.01$, do not reject H_0.

(e) At the 1% level of significance, the data are not significant. The evidence is insufficient to conclude that distribution of dropout rates is different for males and females.

11.

Month	Madison	Juneau	Sign of Difference
1	17.5	22.2	–
2	21.1	27.3	–
3	31.5	31.9	–
4	46.1	38.4	+
5	57.0	46.4	+
6	67.0	52.8	+
7	71.3	55.5	+
8	69.8	54.1	+
9	60.7	49.0	+
10	51.0	41.5	+
11	35.7	32.0	+
12	22.8	26.9	–

(a) $\alpha = 0.05$

H_0: Distributions are the same.

H_1: Distributions are different.

(b) $x = \dfrac{\text{number of plus signs}}{\text{total number of signs}} = \dfrac{8}{12} \approx 0.6667 \approx 0.67$

$$z = \frac{x - 0.5}{\sqrt{\frac{0.25}{n}}} = \frac{0.6667 - 0.5}{\sqrt{\frac{0.25}{12}}} \approx 1.15$$

Use the standard normal distribution.

(c) By Table 5, $P(z > 1.15) = 0.1251$. For a two-tailed test P-value $= 2(0.1251) = 0.2502$.

(d) Since the P-value is greater than the level of significance $\alpha = 0.05$, do not reject H_0.

(e) At the 5% level of significance, the data are not significant. The evidence is insufficient to conclude the temperature distribution in Juneau and Madison are different.

Section 12.2

1.

Yield	Method	Rank
1.10	*B*	1
1.15	*A*	2
1.25	*A*	3
1.34	*B*	4
1.41	*A*	5
1.53	*B*	6
1.61	*A*	7
1.75	*B*	8
1.78	*A*	9
1.80	*B*	10
1.83	*A*	11
1.88	*B*	12
1.89	*B*	13
1.95	*A*	14
1.96	*B*	15
1.99	*A*	16
2.01	*A*	17
2.11	*B*	18
2.12	*A*	19
2.15	*B*	20
2.17	*B*	21
2.21	*B*	22
2.34	*A*	23

(a) $\alpha = 0.05$

H_0: Distributions are the same.

H_1: Distributions are different.

(b) Since $n_1 = 11$ and $n_2 = 12$, R is the sum of the ranks of group A.

$R_A = 2+3+5+7+9+11+14+16+17+19+23 = 126$

$$\mu_R = \frac{n_1(n_1+n_2+1)}{2} = \frac{11(11+12+1)}{2} = 132$$

$$\sigma_R = \sqrt{\frac{n_1 n_2(n_1+n_2+1)}{12}} = \sqrt{\frac{11(12)(11+12+1)}{12}} \approx 16.25$$

$$z = \frac{R-\mu_R}{\sigma_R} = \frac{126-132}{16.25} \approx -0.37$$

Use the standard normal distribution since n_1 and n_2 are each > 10.

(c) By Table 5, $P(z < -0.37) = 0.3557$.
For a two-tail test, P-value = 2(0.3557) = 0.7114.

(d) Since the P-value is greater than the level of significance, do not reject H_0.

(e) At the 5% level of significance, the evidence is insufficient to conclude that the yield distributions are different between organic and conventional farming methods.

3.

Sessions	Group	Rank
19	A	1
20	B	2
24	B	3
25	A	4
26	B	5
27	A	6
28	A	7
31	A	8
33	B	9
34	B	10
35	A	11
36	A	12
37	A	13
38	A	14
39	B	15
40	A	16
41	A	17
42	B	18
43	A	19
44	B	20
45	B	21
46	B	22
48	B	23

(a) $\alpha = 0.05$

H_0: Distributions are the same.

H_1: Distributions are different.

(b) Since $n_1 = 12$ and $n_2 = 11$, R is the sum of the ranks of group B.

$R_B = 2+3+5+9+10+15+18+20+21+22+23 = 148$

$$\mu_R = \frac{n_1(n_1+n_2+1)}{2} = \frac{12(12+11+1)}{2} = 144$$

$$\sigma_R = \sqrt{\frac{n_1 n_2(n_1+n_2+1)}{12}} = \sqrt{\frac{12(11)(12+11+1)}{12}} \approx 16.25$$

$$z = \frac{R-\mu_R}{\sigma_R} = \frac{148-144}{16.25} \approx 0.25$$

Use the standard normal distribution since n_1 and n_2 are each > 10.

(c) By Table 5, $P(z > 0.25) = 0.3974$.
For a two-tail test, P-value = 2(0.3974) = 0.7948.

(d) Since the P-value is greater than the level of significance $\alpha = 0.05$, do not reject H_0.

(e) At the 5% level of significance, the evidence is insufficient to conclude that the distributions of training sessions are different

5.

Minutes	Group	Rank
7	A	1.0
8	A	2.0
9	B	3.0
10	A	4.0
11	A	5.5
11	B	5.5
12	A	7.0
13	A	8.0
14	B	9.0
15	A	10.0
16	A	11.5
16	B	11.5
17	A	13.0
18	A	14.0
19	B	15.0
22	A	16.0
24	B	17.0
27	B	18.0
28	B	19.0
29	B	20.0
30	B	21.0
31	B	22.0
33	B	23.0

(a) $\alpha = 0.05$

H_0: Distributions are the same.

H_1: Distributions are different.

(b) Since $n_1 = 11$ and $n_2 = 12$, R is the sum of the ranks of group A.

$R_A = 1+2+4+5.5+7+8+10+11.5+13+14+16 = 92$

$$\mu_R = \frac{n_1(n_1+n_2+1)}{2} = \frac{11(11+12+1)}{2} = 132$$

$$\sigma_R = \sqrt{\frac{n_1 n_2(n_1+n_2+1)}{12}} = \sqrt{\frac{11(12)(11+12+1)}{12}} \approx 16.25$$

$$z = \frac{R-\mu_R}{\sigma_R} = \frac{92-132}{16.25} \approx -2.46$$

Use the standard normal distribution since n_1 and n_2 are each > 10.

(c) By Table 5, $P(z < -2.46) = 0.0069$.
For a two-tail test, P-value = 2(0.0069) = 0.0138.

(d) Since the P-value is less than the level of significance $\alpha = 0.05$, reject H_0.

(e) At the 5% level of significance, the evidence is sufficient to conclude that the completion time distributions for the two settings are different.

7.

Percent	Group	Rank
29.1	*B*	1
33.7	*B*	2
35.7	*B*	3
36.9	*B*	4
38.2	*B*	5
40.1	*B*	6
44.2	*B*	7
44.3	*A*	8
45.2	*B*	9
46.6	*B*	10
47.1	*A*	11
49.1	*A*	12
50.0	*A*	13
53.3	*B*	14
55.1	*A*	15
57.3	*A*	16
58.2	*A*	17
59.6	*B*	18
59.9	*A*	19
60.0	*A*	20
60.2	*B*	21
63.3	*A*	22
68.7	*A*	23

(a) $\alpha = 0.01$

H_0: Distributions are the same.

H_1: Distributions are different.

(b) Since $n_1 = 11$ and $n_2 = 12$, R is the sum of the ranks of group A.

$R_A = 8+11+12+13+15+16+17+19+20+22+23 = 176$

$$\mu_R = \frac{n_1(n_1+n_2+1)}{2} = \frac{11(11+12+1)}{2} = 132$$

$$\sigma_R = \sqrt{\frac{n_1 n_2(n_1+n_2+1)}{12}} = \sqrt{\frac{11(12)(11+12+1)}{12}} \approx 16.25$$

$$z = \frac{R-\mu_R}{\sigma_R} = \frac{176-132}{16.25} \approx 2.71$$

Use the standard normal distribution since n_1 and n_2 are each > 10.

(c) By Table 5, $P(z > 2.71) = 0.0034$.
For a two-tail test, P-value = 2(0.0034) = 0.0068.

(d) Since the P-value is less than the level of significance $\alpha = 0.01$, reject H_0.

(e) At the 1% level of significance, the evidence is sufficient to conclude that the distributions showing percentage of exercisers differ by education level.

9.

Scores	Group	Rank
61	*A*	1.0
62	*A*	2.0
63	*B*	3.0
65	*A*	4.5
65	*B*	4.5
69	*A*	6.0
70	*B*	7.0
72	*B*	8.0
75	*B*	9.0
77	*A*	10.0
78	*A*	11.0
79	*A*	12.5
79	*B*	12.5
80	*B*	14.0
81	*B*	15.0
82	*B*	16.0
83	*A*	17.0
85	*A*	18.0
87	*A*	19.0
90	*B*	20.0
92	*A*	21.0
93	*A*	22.0
94	*B*	23.0
95	*A*	24.0

(a) $\alpha = 0.01$

H_0: Distributions are the same.

H_1: Distributions are different.

(b) Since $n_1 = 12$ and $n_2 = 12$, R is the sum of the ranks of either group.

$R_A = 1+4.5+6+10+11+12.5+17+18+19+21+22+24 = 166$

$$\mu_R = \frac{n_1(n_1+n_2+1)}{2} = \frac{12(12+12+1)}{2} = 150$$

$$\sigma_R = \sqrt{\frac{n_1 n_2(n_1+n_2+1)}{12}} = \sqrt{\frac{12(12)(12+12+1)}{12}} \approx 17.32$$

$$z = \frac{R-\mu_R}{\sigma_R} = \frac{166-150}{17.32} \approx 0.92$$

Use the standard normal distribution since n_1 and n_2 are each > 10.

(c) By Table 5, $P(z > 0.92) = 0.1788$.
For a two-tail test, P-value = 2(0.1788) = 0.3576.

(d) Since the P-value is greater than the level of significance $\alpha = 0.01$, do not reject H_0.

(e) At the 1% level of significance, the evidence is insufficient to conclude that the distributions of test scores differ according to instruction method.

Section 12.3

1.

Person	Class Rank, x	Sales Rank, y	$d = x - y$	d^2
1	6	4	2	4
2	8	9	−1	1
3	11	10	1	1
4	2	1	1	1
5	5	6	−1	1
6	7	7	0	0
7	3	8	−5	25
8	9	11	−2	4
9	1	3	−2	4
10	10	5	5	25
11	4	2	2	4
Sum	66	66	0	70

(a) $\alpha = 0.05$

H_0: $\rho_s = 0$ (there is no monotone relationship between x and y)

H_1: $\rho_s \neq 0$ (there is a monotonic relationship between x and y)

(b) $$r_s = 1 - \frac{6\Sigma d^2}{n(n^2-1)} = 1 - \frac{6(70)}{11(11^2-1)} \approx 0.682$$

(c) From Table 9, 0.682 falls between entries 0.619 and 0.764 in the n = 11 row. Use two-tailed areas to find that 0.01 < P-value < 0.05.

(d) Since the P-value is less than the level of significance $\alpha = 0.05$, reject H_0.

(e) At the 5% level of significance, we conclude that there is a monotonic relationship (either increasing or decreasing) between the rank in training class and the rank in sales.

3.

Rat Colony	Population Density Rank, x	Violence Rank, y	$d = x - y$	d^2
1	3	1	2	4
2	5	3	2	4
3	6	5	1	1
4	1	2	−1	1
5	8	8	0	0
6	7	6	1	1
7	4	4	0	0
8	2	7	−5	25
Sum	36	36	0	36

(a) $\alpha = 0.05$

H_0: $\rho_s = 0$ (there is no monotonic relationship)

H_1: $\rho_s > 0$ (the relationship between x and y is monotone increasing; the higher population density ranks are associated with the higher violence ranks)

(b) $r_s = 1 - \dfrac{6\sum d^2}{n(n^2-1)} = 1 - \dfrac{6(36)}{8(64-1)} = 0.571.$

(c) From Table 9, 0.571 falls to the left of the entry 0.620 in the $n = 8$ row. Use one-tail areas to find that P-value > 0.05.

(d) Since the P-value is greater than the level of significance $\alpha = 0.05$, do not reject H_0.

(e) At the 5% level of significance, there is insufficient evidence to indicate a monotonic-increasing relationship between crowding and violence.

5. (i)

Soldier	Humor Test Score	Humor Test Rank, x	Aggressiveness Test Score	Aggressiveness Rank, y	$d = x - y$	d^2
1	60	5	78	1	4	16
2	85	3	42	7	−4	16
3	78	4	68	3	1	1
4	90	2	53	5	−3	9
5	93	1	62	4	−3	9
6	45	7	50	6	1	9
7	51	6	76	2	4	1
Sum	-	28	-	28	0	68

The exact same answer will result with rank 1 assigned to the largest value or the smallest value.

(ii) (a) $\alpha = 0.05$

H_0: $\rho_s = 0$ (there is no monotonic relationship)

H_1: $\rho_s < 0$ (there is a monotone-decreasing relationship between x and y)(here, soldiers with a greater sense of humor, (smaller rank number) have lower aggression scores (larger rank number))

(b) $r_s = 1 - \dfrac{6\sum d^2}{n(n^2-1)} = 1 - \dfrac{6(68)}{7(48)} \approx -0.214,$

(c) From Table 9, 0.214 falls to the left of entry 0.715 in the $n = 7$ row. Use the one-tailed areas to find that P-value > 0.05.

(d) Since the P-value is greater than the level of significance $\alpha = 0.05$, do not reject H_0.

(e) At the 5% level of significance, the evidence is insufficient to conclude that there is a monotonic decreasing relationship between ranks of humor and ranks of aggressiveness.

7. (i)

Area	Police Rank, x	Fire Fighter Rank, y	$d = x - y$	d^2
1	12	12.0	0.0	0.00
2	7	8.0	−1.0	1.00
3	10	10.0	0.0	0.00
4	5	4.0	1.0	1.00
5	4	5.5	−1.5	2.25
6	2	2.0	0.0	0.00
7	6	5.5	0.5	0.25
8	9	11.0	−2.0	4.00
9	3	3.0	0.0	0.00
10	13	13.0	0.0	0.00
11	11	7.0	4.0	16.00
12	8	9.0	−1.0	1.00
13	1	1.0	0.0	0.00
Sum	91	91	0	25.5

The exact same answer will result with rank 1 assigned to the largest value or the smallest value.

(ii) (a) $\alpha = 0.05$

H_0: $\rho_s = 0$ (no monotone relationship)

H_1: $\rho_s \neq 0$ (monotone relationship)

(b) $$r_s = 1 - \frac{6\Sigma d^2}{n(n^2 - 1)} = 1 - \frac{6(25.5)}{13(13^2 - 1)} \approx 0.930$$

(c) From Table 9, 0.930 falls to the right of the entry 0.797 in the $n = 13$ row. Use the two-tailed areas to find that P-value < 0.002.

(d) Since the P-value is less than the level of significance $\alpha = 0.05$, reject H_0.

(e) At the 5% level of significance, we conclude that there is a monotonic relationship between number of fire fighters and number of police.

9. (i)

City	Insurance Sales Rank, x	Per Capita Income	Income Rank, y	$d = x - y$	d^2
1	6	17	5	1	1
2	7	18	4	3	9
3	1	19	2.5	−1.5	2.25
4	8	11	8	0	0
5	3	16	6	−3	9
6	2	20	1	1	1
7	5	15	7	−2	4
8	4	19	2.5	1.5	2.25
Sum	36	-	36	0	28.5

The exact same answer will result with rank 1 assigned to the largest value or the smallest value.

(ii) (a) $\alpha = 0.01$

H_0: $\rho_s = 0$ (there is no monotonic relationship)

H_1: $\rho_s \neq 0$ (there is a monotone relationship between x and y)

(b) $r_s = 1 - \dfrac{6\Sigma d^2}{n(n^2-1)} = 1 - \dfrac{6(28.5)}{8(64-1)} \approx 0.661$

(c) From Table 9, 0.661 falls between entries 0.620 and 0.715 in the $n = 8$ row. Use two-tailed areas to find that $0.05 < P\text{-value} < 0.10$.

(d) Since the P-value is greater than the level of significance $\alpha = 0.01$, do not reject H_0.

(e) At the 1% level of significance, we conclude that there is insufficient evidence to reject the null hypothesis of no monotonic relationship between rank of insurance sales and rank of per capita income.

Section 12.4

1. (a) $\alpha = 0.05$;

H_0: The symbols are randomly mixed in the sequence.

H_1: The symbols are not randomly mixed in the sequence.

(b) RRR|DD|RR|DDDDD|R|D|RR|D|RRR|DD|R

$R = 11$

(c) Letting R be the first symbol, $n_1 = 12$ and $n_2 = 11$. From Table 10, $c_1 = 7$ and $c_2 = 18$.

(d)

$R \leq 7$	$8 \leq R \leq 17$	$R \geq 18$
Reject H_0	Fail to reject H_0	Reject H_0

Since $R = 11$, do not reject H_0.

(e) At the 5% level of significance, the evidence is insufficient to conclude that the sequence of presidential party affiliation is not random.

3. (a) $\alpha = 0.05$;

H_0: The symbols are randomly mixed in the sequence.

H_1: The symbols are not randomly mixed in the sequence.

(b) SSS|N|S|N|SSSS|NN|S|N|SSS|NN|SSSS

$R = 11$

(c) Letting S be the first symbol, $n_1 = 16$ and $n_2 = 7$. From Table 10, $c_1 = 6$ and $c_2 = 16$.

(d)

$R \le 6$	$7 \le R \le 15$	$R \ge 16$
Reject H_0	Fail to reject H_0	Reject H_0

Since $R = 11$, we do not reject H_0.

(e) At the 5% level of significance, the evidence is insufficient to conclude that the sequence of days for seeding and not seeding is not random.

5. (i) Median = 11.7; using A for above the median and B for below, the original sequence translates to BBBAAAAABBBA

(ii) (a) $\alpha = 0.05$;

H_0: The numbers are randomly mixed about the median.

H_1: The numbers are not randomly mixed about the median.

(b) BBB|AAAAA|BBB|A

$R = 4$

(c) Letting B be the first symbol, $n_1 = 6$ and $n_2 = 6$. From Table 10, $c_1 = 3$ and $c_2 = 11$.

(d)

$R \le 3$	$4 \le R \le 10$	$R \ge 11$
Reject H_0	Fail to reject H_0	Reject H_0

Since $R = 4$, do not reject H_0.

(e) At the 5% level of significance, the evidence is insufficient to conclude that the sequence of returns is not random about the median.

7. (i) Median = 21.6; using A for above the sequence and B for below, the original sequence translates to BAAAAAABBBBB

(ii) (a) $\alpha = 0.05$;

H_0: The numbers are randomly mixed about the median.

H_1: The numbers are not randomly mixed about the median.

(b) B|AAAAAA|BBBBB

$R = 3$

(c) Letting B be the first symbol, $n_1 = 6$ and $n_2 = 6$. From Table 10, $c_1 = 3$ and $c_2 = 11$.

(d)

$R \le 3$	$4 \le R \le 10$	$R \ge 11$
Reject H_0	Fail to reject H_0	Reject H_0

Since $R = 3$, reject H_0.

(e) At the 5% level of significance, we can conclude that the sequence of percentage of sand in the soil at successive depths is not random about the median.

9. (a) H_0: The symbols are randomly mixed in the sequence.

H_1: The symbols are not randomly mixed in the sequence.

(b) $n_1 = 21;\ n_2 = 17;\ R = 18$

(c) $\mu_R = \dfrac{2n_1n_2}{n_1+n_2} + 1 = \dfrac{2(21)(17)}{21+17} + 1 \approx 19.80$

$$\sigma_R = \sqrt{\frac{(2n_1n_2)(2n_1n_2 - n_1 - n_2)}{(n_1+n_2)^2(n_1+n_2-1)}} = \sqrt{\frac{2(21)(17)[2(21)(17) - 21 - 17]}{(21+17)^2(21+17-1)}} \approx 3.01$$

$$z = \frac{R - \mu_R}{\sigma_R} = \frac{18 - 19.80}{3.01} \approx -0.60$$

(d) Since $-1.96 < -0.60 < 1.96$, do not reject H_0. $P(z < -0.60) = 0.2743$; P-value $\approx 2(0.2743) = 0.5486$; At the 5% level of significance, the P-value also tells us not to reject H_0.

(e) At the 5% level of significance, the evidence is insufficient to reject the null hypothesis of a random sequence of Democrat and Republican presidential terms.

Chapter 12 Review

1. (a) Rank-sum test

Index	Group	Rank
1.1	1	1
1.5	2	2
1.6	1	3
1.8	1	4
1.9	2	5
2.2	2	6
2.4	2	7
2.5	1	8
2.8	2	9
2.9	1	10
3.1	2	11
3.2	1	12
3.3	2	13
3.5	2	14
3.6	2	15
3.7	1	16
3.8	1	17
3.9	2	18
4.0	2	19
4.1	1	20
4.2	1	21
4.4	1	22
4.6	2	23

(b) $\alpha = 0.05$

H_0: Distributions are the same.

H_1: Distributions are different.

(c) Since $n_1 = 11$ and $n_2 = 12$, R is the sum of the ranks of group 1.

$R_1 = 1+3+4+8+10+12+16+17+20+21+22 = 134$

$$\mu_R = \frac{n_1(n_1+n_2+1)}{2} = \frac{11(11+12+1)}{2} = 132$$

$$\sigma_R = \sqrt{\frac{n_1 n_2(n_1+n_2+1)}{12}} = \sqrt{\frac{11(12)(11+12+1)}{12}} \approx 16.25$$

$$z = \frac{R-\mu_R}{\sigma_R} = \frac{134-132}{16.25} \approx 0.12$$

(d) By Table 5, $P(z > 0.12) = 0.4522$.
For a two-tail test, P-value $= 2(0.4522) = 0.9044$.

(e) Since the P-value is greater than the level of significance $\alpha = 0.05$, do not reject H_0. At the 5% level of significance, there is insufficient evidence to conclude that the viscosity index distribution has changed with the catalyst.

3. (a) Sign Test

Sales After	Sales Before	Sign of Difference
610	460	+
150	216	–
790	640	+
288	250	+
715	685	+
465	430	+
280	220	+
640	470	+
500	370	+
118	118	N.D.
265	117	+
365	360	+
93	93	N.D.
217	291	–
280	430	–

(b) $\alpha = 0.01$

H_0: Distributions are the same.

H_1: Distributions after ads is higher.

(c) $$x = \frac{\text{number of plus signs}}{\text{total number of signs}} = \frac{10}{13} \approx 0.77$$

$$z = \frac{x-0.5}{\sqrt{\frac{0.25}{n}}} = \frac{0.77-0.5}{\sqrt{\frac{0.25}{13}}} \approx 1.95$$

Use the standard normal distribution.

(d) By Table 5, the area in the right tail is $P(z > 1.95) = 0.0256$; P-value $= 0.0256$.

(e) Since the P-value is greater than the level of significance $\alpha = 0.01$, do not reject H_0.

At the 1% level of significance, the evidence is insufficient to claim that the distribution is higher after the ads.

5.

Employee	Training Program Rank, x	Rank on the Job, y	$d = x - y$	d^2
1	8	9	−1	1
2	9	8	1	1
3	7	6	1	1
4	3	7	−4	16
5	6	5	1	1
6	4	1	3	9
7	1	3	−2	4
8	2	4	−2	4
9	5	2	3	9
Sum	45	45	0	46

(a) Spearman rank correlation coefficient test.

(b) $\alpha = 0.05$
H_0: $\rho_s = 0$ (there is no monotonic relationship)
H_1: $\rho_s > 0$ (there is a monotone-increasing relationship)

(c) $r_s = 1 - \dfrac{6\Sigma d^2}{n(n^2-1)} = 1 - \dfrac{6(46)}{9(81-1)} \approx 0.617$

(d) From Table 9, 0.617 falls between entries 0.600 and 0.700 in row $n = 9$. Use one-tailed areas to find that $0.025 < P\text{-value} < 0.05$.

(e) Since the P-value is less than the level of significance $\alpha = 0.05$, reject H_0.
At the 5% level of significance, we conclude there is a monotone-increasing relation in the ranks from the training program and from the job.

7. **(a)** Runs test for randomness.

(b) $\alpha = 0.05$
H_0: The symbols are randomly mixed in the sequence.
H_1: The symbols are not randomly mixed in the sequence.

(c) TTTT|F|TT|FF|TTTT|FFFFFF|TTTTTT
$R = 7$

(d) Letting T be the first symbol, $n_1 = 16$ and $n_2 = 9$.
From Table 10, $c_1 = 7$ and $c_2 = 18$.

(e)

$R \le 7$	$8 \le R \le 17$	$R \ge 18$
Reject H_0.	Fail to reject H_0.	Reject H_0.

Since $R = 7$, we reject H_0.
At the 5% level of significance, we conclude that the sequence of answers is not random.

NOTES

NOTES

NOTES

NOTES

NOTES

NOTES

NOTES

NOTES

NOTES

NOTES

NOTES

NOTES